Accounting for Construction

Accounting for Construction follows on from *Measuring Construction*, edited by the same team. It extends the coverage of some of the material in the first volume and expands the range of related topics to include, *inter alia*, shadow economies, accounting for informal construction and the treatment of the built environment sector in national accounts.

Taken together, the two volumes collate a range of topics that are only addressed, if addressed at all, in occasional academic papers and the publications of bodies such as national statistical offices and the World Bank. *Accounting for Construction* presents international examples from the UK, Australia and New Zealand and from both academic and professional contributors.

This book is essential reading for researchers and professionals interested in construction economics, construction management, and anyone interested in how the construction industry affects the global economy in ways previously under-represented in the literature.

Rick Best is Associate Professor of Construction Management and Economics at Bond University. He has produced numerous book chapters and papers over a 25-year career as an academic, as well as co-editing four books and co-authoring one quantity surveying textbook. He was the founding director of the Centre for Comparative Construction Research. His research over the past 10 years has been focused on the problems associated with making valid comparisons of construction industries across countries and has contributed to the development of a construction data collection project within the International Comparison Program.

Jim Meikle has a part-time chair in the economics of the construction sector at the Bartlett School of Construction and Project Management, University College London, and is an adjunct professor at Bond University. He retired as a partner of Davis Langdon LLP in 2005, and since then he has worked as an independent consultant with UK government departments and agencies, foreign governments, international organizations including the EU, the World Bank and the African Development Bank, and private clients. His main interests are international comparisons, construction industry policy, construction data, construction productivity and construction professional services.

Accounting for Construction

Frameworks, Productivity, Cost
and Performance

Edited by Rick Best and Jim Meikle

Routledge
Taylor & Francis Group

LONDON AND NEW YORK

First published 2019
by Routledge
2 Park Square, Milton Park, Abingdon, Oxon OX14 4RN

and by Routledge
605 Third Avenue, New York, NY 10017

First issued in paperback 2021

Routledge is an imprint of the Taylor & Francis Group, an informa business

British Library Cataloguing-in-Publication Data
A catalogue record for this book is available from the British Library

Library of Congress Cataloging-in-Publication Data
Names: Best, Rick, author. | Meikle, Jim, author.
Title. Accounting for construction ; frameworks, productivity, cost and
    performance / [edited by] Rick Best & Jim Meikle.
Description: Abingdon, Oxon ; New York, NY : Routledge, 2019. |
    Includes bibliographical references.
Identifiers: LCCN 2018054013 | ISBN 9781138293977 (hbk) |
    ISBN 9781315231785 (ebk)
Subjects: LCSH: Building—Cost effectiveness. | Building—Estimates. |
    Construction industry—Accounting.
Classification: LCC TH438.15 .A24 2019 | DDC 657/.869—dc23
LC record available at https://lccn.loc.gov/2018054013

ISBN 13: 978-1-03-209324-6 (pbk)
ISBN 13: 978-1-138-29397-7 (hbk)

Typeset in Goudy
by Apex CoVantage, LLC

In memory of Michael Regan

Contents

Contributors

Malcolm Abbott is Associate Professor of Economics at the Swinburne University of Technology in Melbourne and has a PhD from the University of Melbourne. One of his main areas of expertise is productivity analysis, and he has published extensively on the productivity of utilities (electricity and gas), hospitals, airports and the construction industry in Australia and New Zealand. His other fields of research include the development of energy markets, utility regulation and the economics of education.

Rick Best is Associate Professor of Construction Management and Economics at Bond University. He has produced numerous book chapters and papers over a 25-year career as an academic, as well as co-editing four books and co-authoring one quantity surveying textbook. He was the founding director of the Centre for Comparative Construction Research. His research over the past 10 years has been focused on the problems associated with making valid comparisons of construction industries across countries and has contributed to the development of the most recent construction data collection project within the International Comparison Program.

Chris Carson is a lecturer and previous Programme Leader in the Bachelor of Construction program at Unitec Institute of Technology, Auckland, New Zealand. With an MA in Economics, he is responsible for teaching the Economics and third year Property Courses in the Bachelor of Construction. He has contributed research into the area of productivity in the New Zealand construction industry and is currently editing an encyclopaedia of contract administration. His other research interests lie in the history of economic thought.

Toong Khuan Chan is Assistant Dean (International) and Senior Lecturer in the Faculty of Architecture, Building and Planning at the University of Melbourne. His research interest is currently focused on the nexus of economics and building technology with emphasis on issues of industry development, productivity, competitiveness and profitability. He has received funding from industry and government to support research into firm and industry performance, globalization of construction supply chains, intelligent transportation systems, wireless sensor networks and concrete structures.

Will Chancellor has a PhD from Swinburne University of Technology and is an economist at the Australian Bureau of Agricultural and Resource Economics and Sciences (ABARES) in Canberra. His research has focused on construction industry productivity and efficiency analysis, extending into international case studies and novel topics such as the distorting effect of the shadow economy on construction productivity. Other research interests include residential property prices, productivity drivers and agricultural productivity.

David Chandler OAM is the Principal of CE Advisory, offering a specialist service to enterprises, government and investors who seek practical insight and strategic advice in the modern construction world. He was the construction director for Australia's parliament building and has since led enterprises involved with a range of major projects. He is Adjunct Professor and Industry Engagement Lead for the Centre for Smart Modern Construction (c4SMC) at Western Sydney University. The Centre involves an industry and academic collaboration to invest in modern construction research and academic capability building to help prepare the next generation of constructors to be future ready.

D'Maris Coffman is the Professor of Economics and Finance of the Built Environment at the Bartlett Faculty of the Built Environment, University College London and the Director of the Bartlett School of Construction and Project Management. She publishes widely, primarily in financial history, infrastructure economics and climate change economics and finance. She is the Senior Editor of Palgrave *Studies in the History of Finance*, Managing Editor of *Structural Change and Economic Dynamics* and Associate Editor of *Economia Politica*.

Gerard de Valence is a senior lecturer in the School of the Built Environment at the University of Technology Sydney. He also has a visiting appointment at the Bartlett School of Construction and Project Management, University College London. Prior to becoming an academic in 1992, he worked in the private sector doing research on the property, building and construction industries, and worked on industry policy in the 1990s with Australian State and Commonwealth Governments. His research has focused on issues around the structure, conduct and technological trajectory of the construction industry. He has edited and contributed to a number of books, published numerous papers and was coordinator of the CIB Working Commission on Building Economics (W55) between 2003 and 2011.

Gary Emmett is Senior Economist for Turner & Townsend, undertaking consulting assignments for private sector and government clients and providing economic advice to the Turner & Townsend business. He researches construction markets, analyzes data and prepares benchmark information to support large construction projects. He has expertise in cost movements of major capital expense items and labour, both historic and forecast. He prepares cost escalation indices (historic and forecast). He researches and authors the annual International Construction Market Survey report, now in its tenth year. Gary has an MA Economics from the University of Edinburgh and Graduate Certificate in Property Economics from Queensland University of Technology.

Brian Green is an independent researcher and commentator on the construction and housing sectors. A former journalist, leaving *Construction News* as Managing Editor in 1998, he maintains a wide interest in the built environment, blogging as *Brickonomics* and writing regularly on economic trends for *RIBA Journal*. He is editor of the Housing Market Intelligence report associated with the Home Builders Federation and NHBC annual conference. In recent years he has worked closely with the Chartered Institute of Building, producing reports on topics as varied as productivity, migration and social mobility. He is an honorary research associate at the Bartlett School of Construction and Project Management, University College London.

Mary Hardie is Associate Professor in Construction Management at Western Sydney University. She has published extensively in refereed journals and produced book chapters and numerous conference papers over the past 15 years. Her research focuses on technical innovation in the construction industry, and its generation and diffusion among SMEs. Her secondary research area is construction education. She is a founding member of the Centre for Smart Modern Construction (c4SMC) at Western Sydney University. The Centre seeks to address the challenges facing the Australian construction industry in an age of rapidly changing technology, digitalization and globalization.

John Kelsey is Lecturer in Construction, Project Management and Economics at the School of Construction and Project Management in the Bartlett Faculty of the Built Environment at University College London. He has wide-ranging industrial experience in the accounting and commercial aspects of housing, construction, civil engineering and mining. He was a Senior Cost Engineer on the Channel Tunnel Project 1989–1994. His research has been mainly in the fields of digital support for construction planning, management and stakeholder engagement together with technology transfer and the impact of culture and organizational learning on health and safety management. He is currently course leader for the MSc Construction Economics and Management at UCL.

Craig Langston is Professor of Construction and Facilities Management at Bond University in Queensland. In addition to authoring more than 100 papers, five books and three software programs, Professor Langston has been awarded four Australian Research Council (ARC) grants totalling about AUD 1 million, won the Chartered Institute of Building International Innovation and Research Award (2013), Outstanding Paper for the *Facilities* journal (2013), Emerald Literati Network Outstanding Paper Award (2013) and Bond University Vice Chancellor's Quality Award for Research Excellence (2010). He has 30 years' experience as an academic, teaching a range of subjects related to the built environment. He is also Director of the Centre for Comparative Construction Research at Bond.

Jim Meikle has a part-time chair in the economics of the construction sector at the Bartlett School of Construction and Project Management, University College London, and is an adjunct professor at Bond University. He retired as a partner of Davis Langdon LLP in 2005, and since then he has worked as an independent consultant with UK government departments and agencies, foreign governments, international organisations including the EU, the World Bank and the African Development Bank, and private clients. His main interests are international comparisons, construction industry policy, construction data, construction productivity and construction professional services.

Srinath Perera is the Director of c4SMC and holds a personal chair in Built Environment and Construction Management at the Western Sydney University, Australia. He is a coordinator of the CIB TG83 Task Group, e-Business in Construction. He is a Fellow of the Royal Society of New South Wales and is a fellow of the Australian Institute of Building (AIB). He is a chartered quantity surveyor and project manager. He has extensive experience in doctoral student supervisions and examinations and has over 150 peer-reviewed publications. He is co-author of the popular textbooks *Cost Studies of Buildings* (6th ed.), *Contractual Procedures in the Construction Industry* (7th ed.) and *Advances in Construction ICT & e-Business*, published by Routledge.

Michael Regan was Professor of Infrastructure at Bond University and had qualifications in law, economics, finance and infrastructure economics. His research work concerned infrastructure investment, financing and operational efficiency in Asia with his research widely published internationally. Michael was also technical adviser to the Economist Intelligence Unit 2018 Asian Infrascope Report and undertook research studies for the Asian Development Bank, the United Nations Economic and Social Commission for East Asia, and the Economic Research Unit for ASEAN and East Asia (ERIA). Recent reports include the ASEAN PPP Manual and an institutional study for joint sponsors, Asian Development Bank Institute, the OECD and ERIA. Sadly, Michael passed away while this book was in the final stages of production. The editors dedicate this book to Michael and remember him as a fine academic, a true professional and a wonderful colleague.

Preface

This book is the second in a series that explores a range of topics related to construction projects, firms and the industry as a whole, at both the national and international levels. The first in the series, *Measuring Construction*, explored matters related to measuring industry output, productivity and costs/prices; in this second volume the scope broadens to include several chapters on construction in national accounts as well as presenting further thoughts on industry comparisons, construction productivity assessment and a range of other interrelated topics. For example, the lack of consistency in the way that construction industry data is collected and how it is aggregated and/or disaggregated before becoming made publicly available is a major hurdle for academics and other researchers in government agencies and private consultancies. This issue was raised in the first book and is an important thread that runs through several chapters here. Similarly, the issue of construction productivity assessment is further explored with, for example, discussion of the effects on construction productivity that are the result of changes in the mix of labour and capital inputs. A subsequent chapter investigates technology choices in developing economies that looks at the tensions between a plentiful supply of relatively cheap labour and the adoption of more capital-intensive technologies and processes that may or may not be the most appropriate choices where employment for low-skilled or unskilled workers and capital for investment in exotic machines and materials are both limited.

In several instances, individual authors put forward suggestions for different ways to address specific issues; in Chapter 2, for example, an alternative approach to the categorization of construction industry data in national accounts is proposed. Two chapters review past exercises in international cost comparisons while others cover diverse topics such as the potential use of 'big data' by construction firms and methods for the estimation of the extent of the cash or shadow economy in the context of national accounts.

We hope that this second volume will be a useful addition to the literature of construction economics and help students and practitioners to better understand some of the complexities of the construction industry that both intrigue and frustrate us in more or less equal measure.

Note: In this book, as in its companion volume, *Measuring Construction*, the word 'data' is treated as a collective noun and thus it takes the singular form as in '. . . the data is reliable . . .' rather than the somewhat archaic '. . . the data are reliable. . . .' While in Latin the singular is 'datum' and the plural 'data', this book is written in English, not Latin, and the editors are not aware of any common usage of the word 'datum' to describe or identify a single piece of data.

Rick Best – Gold Coast
Jim Meikle – Geelong
October 2018

1 The challenges of measuring and accounting for construction

Rick Best and Jim Meikle

Introduction

Measuring construction used to be a straightforward exercise. Work was physically measured on completion and those who did the work were paid based on the quantities of work measured. People who carried out the measurement or 'surveying' work became known as quantity surveyors.

Gradually the practice of measuring and estimating the cost of construction before the start of the work, usually from some sort of drawing(s), replaced the measurement of work after it was completed. A handwritten estimate for the building of a cottage in Wales prepared in 1809 (see Lethbridge 2008) included the following items:

- For digging stones for building 294 yds @ 8d = £ 9.16.0
- For building the house   294 yds @ 14d = £ 17.3.0
- 220 feet of timber @ 4/6 per foot   = £49.10.0
- All sorts of nails   = £ 3. 0.0
- Hinges, latchets and smyth's work   = £ 0.15.0

The entire estimate comprised just 14 items. In Commonwealth countries at least, such measurement and estimation developed into the detailed measurement and pricing of building works, with measurement based on precise rules compiled and published by professional bodies such as the Royal Institution of Chartered Surveyors (RICS).

This sort of approach, with variations, has worked well and is still used, albeit with quantities now being extracted from digital 3D building models and linked to databases containing unit rates for specific work items. In this book, however, and its precursor, *Measuring Construction* (Best and Meikle 2015), the focus is not on the measurement of work items. In these books a range of more challenging questions is explored; they relate to the measurement of different aspects of construction (such as productivity), at project level, at the firm level, at the level of national industries and even the measurement of construction as a whole, as a worldwide affair. That includes how the creation of the built environment (which includes infrastructure such as roads and bridges as well as buildings of all shapes and sizes) is accounted for by national statistical agencies.

Accounting for construction at these broader levels requires measurement, and that takes many forms. Each presents its own set of challenges. Before exploring some aspects of the larger problems in more detail, it is worth revisiting the need for such measurement.

Why measure?

It is commonly held that measurement is an essential part of management and the adage 'if you can't measure it, you can't manage it' is often repeated in support of this notion. Actually, the quote is incomplete and misleading. It should be: 'It is wrong to suppose that if you can't measure it, you can't manage it – a costly myth' (Deming 1994). Deming was talking about managing things that could not be measured in the context of business improvement. Harrington (1987), also in the context of business improvement, linked measurement to understanding, and suggested that measurement enabled understanding, understanding made it possible to control, and control could then lead to improvement. The search for understanding is often the primary reason for the sort of measurement discussed here; for example, understanding differences in how buildings are measured, or how different parts of the construction industry are accounted for in the national statistics of different countries. It is measurement that provides the information that enables governments and their agencies to account for construction in the context of economic activity and whole economies, and it is the understanding that is achieved that enables researchers and others to more correctly interpret the statistics related to building and construction activity.

In some instances, the key driver is the intent to draw comparisons. Sometimes this is in the context of performance improvement (e.g. increasing productivity), and in other cases it is more about establishing metrics that can provide reliable data to decision-makers (e.g. real gross domestic product [GDP] per capita – see ICP 2015). In most cases there is the need to measure some aspect or aspects of the industry (e.g. productivity, cost, quality, time to complete), or some combination of such factors. While construction is often described as 'comparison-resistant' (see, for example, Meikle and Thomas 2012), many attempts at comparisons of cost and productivity (*inter alia*) have been made, and there is ongoing research in this area. Measurement of some sort is invariably at the heart of any attempt to compare aspects or segments of the industry or to compare whole industries. Equally, it is industry measurements of varying types that define and describe construction in national accounts.

Defining an industry

One fundamental accounting concern is how we define the construction industry and thus how we define what it is that we are measuring, and how we include it in national accounts. Is it restricted to firms that are actively engaged in physically constructing projects, or should it include, for example, suppliers of materials and components, and designers and managers of construction projects?

Building work undertaken as part of maintenance, renovation and refurbishment is an added complication; in many places this sort of work represents a large part of total construction activity, but its value may or may not be included in measures of construction industry activity.

Informal construction, part of the so-called shadow economy, is building activity that is often unrecorded but may represent a significant proportion of the total, particularly in developing economies where building work is often done by households. Those who assist are often unpaid or are paid in cash or in kind; in either case the value is often under-reported and/or under-recorded.

Even the exact nature of the industry is not clear; it has some similarities to manufacturing while having some unique characteristics that set it apart from manufacturing, the most notable being the heterogeneity of its products. The problem of what constitutes the industry makes even an apparently simple measure, such as construction's contribution to a country's GDP, a tricky question that requires more than just a number to describe it adequately. National accounts in different countries measure construction activity in different ways, and that adds further layers of complexity to the challenges of comparing industries or even specific industry characteristics across national boundaries.

Construction is not, in fact, a single industry; it is a group of economic activities that use a more or less common set of resources to produce built assets. Those activities are grouped together as an industry in the Standard Industrial Classification (SIC) but the differences in both process and product between, for example, small domestic repair and maintenance (R&M) and large infrastructure projects are very large. In contrast, the SIC definition for the car industry, for example, excludes R&M.

Construction in national accounts

Official data on construction activity can be confusing. Typically there are at least two main types of data: contractors' construction output and construction output in the national accounts. Brian Green's chapter describes in detail the measurement and presentation of contractors' construction output in the UK, only one country but illustrative of many; and the chapter illustrates how complicated that can be.

Construction in national accounts can be even more complicated. It can be presented as value added – the construction industry's contribution to construction output – or gross output – construction output including works and services bought in from other industries. Value added data is the basis of production and income versions of the national accounts; gross output is the basis of the expenditure version. And measures of construction activity can comprise all construction activity or only formal construction output, economic activity by firms registered to construction in the SIC. This latter may include contracting firms' activities in addition to construction work (property development or building materials manufacture, for example). Construction in the national accounts should also include work by informal/unregistered firms or individuals and capital works (new works and major improvements) by households, but this is not always the case. Construction work undertaken in-house by organisations – direct labour or own account – is typically included as output by their industry; construction work undertaken by employees of a chemical products company is, therefore, chemical industry, not construction industry, output.

Construction professional services (architecture, engineering, project management) may be included as construction output or may be included as professional services. Partly this may be a result of how construction activity is defined in a country (see Meikle and Grilli 1999) or just because the boundaries between construction and professional services are blurred. A number of procurement approaches – Design and Build or Public Private Partnerships, for example – include a range of works and services in their contract values and it is often difficult, if not impossible, to separate expenditures. Some procurement approaches include technological components – signalling in rail projects or wind turbines in alternative energy projects – and again, their costs can be difficult to separate from construction work and may end up included as construction output.

The difficulties associated with the measurement of construction in the national accounts are important and deserving of further study.

The volume of output

One key difficulty lies in how we measure the quantity or *volume* of construction output. It is not necessarily an easy concept to grasp. Expressing output volume in terms of, say, floor area of residential construction does not provide a particularly useful quantity, as even if high-rise and other multi-unit residential construction is excluded, there is still considerable variety in the nature of the output as it includes everything from modest cottages for low-income families to lavish mansions for the well-to-do. Differences in the 'typical family home' between different regions adds greatly to the complexity – compare, for example, the Nordic house and the Portuguese house that are part of the Eurostat-OECD price comparison exercise (Eurostat-OECD 2012). Comparing total floor area of housing in one place, or one time period, with that in another also ignores differences in scale, quality, complexity, materials and standard (quality) of fitout.

The heterogeneity of the industry's output leaves us with little alternative other than to express volume in terms of monetary value, yet this method has some obvious shortcomings. The amount paid for work done also offers little information or insight into the nature of the resultant output. For example, is $1 million worth of factory the same *amount* of construction as $1 million worth of prestige office? Both are 'non-residential construction', but the two products are physically very different, and the costs are markedly different. In Sydney in 2017, AUD 1 million represented 500–1000 m^2 of factory (depending on inclusions such as showrooms and offices) compared to around 200 m^2 of fully fitted prestige office (Rawlinsons 2017).

The volume of output can be expressed in other ways; in *Measuring Construction* the idea of purchasing power parity (PPP) was described and construction-specific applications of the approach were explored. One example was based on the Big Mac Index; construction costs expressed in any national currency can be divided by the cost of a Big Mac hamburger in the same currency, and thus the cost of construction can be expressed as the number of hamburgers per square metre, or total cost can be measured in hamburgers, with the hamburger as an artificial unit of 'currency'. Buildings are not, however, constructed of hamburgers so the valid-ity of this method is questionable[1] (see, for example, Croce *et al.* 1999).

The underlying purpose of PPPs is to eliminate price level differences between different locations. General PPPs are useful when a person is contemplating tak-ing a job in another country where they will be paid in that country's currency. Simply converting the potential salary in a foreign currency to the equivalent in the home currency, using exchange rates, provides little insight into whether the potential salary is better than, or about the same as, the person's current salary in terms of how well they can live on the salary being offered compared to their present lifestyle. The important point is how much can the person buy in the new country with the new salary compared to what they can buy with their current salary in their home country; that is, how will the purchasing power of the new salary compare to their current purchasing power?

'How much' the person can buy depends, of course, on what they buy. For someone relocating to another country, the primary concerns will probably be living expenses: rent, food, energy, clothing and so on. General PPPs that are based on the costs of a large range of goods and services may be useful, and the World Bank's general or GDP PPPs are computed using data on the cost of lots of items that are not relevant when simply comparing one salary with another – heavy machinery for example, as these PPPs are used to compare economic indi-cators on a larger scale, such as GDP per capita – but detailed PPPs are available for things like clothing, food and beverages, and housing.

The problem has been addressed in another way by using a representative bas-ket of typical inputs to construction projects (including items of labour, plant and materials). The basic approach involves the pricing of a set of inputs that are considered to be reasonably representative of construction so that the basket that is priced becomes, in effect, a unit of construction currency. The cost of construction can then be expressed as the number of baskets per square metre, or total project cost can be expressed as a number of baskets. Unlike the Big Mac, such a basket consists of real inputs to the construction process and are thus more representative of the industry's output. The construction component of the Inter-national Comparison Program (ICP) uses a variant of this approach, and that exercise is described in some detail in *Measuring Construction* (see Chapter 4) and later in this book (see Chapter 9). Another variant is discussed by Langston in Chapters 8 and 9 of *Measuring Construction* and a practical application described by Emmett and Langston in Chapter 10 here.

Alternatively, the cost of baskets priced in different locations can be compared and PPP factors calculated that can be used to compare construction costs recorded in different currencies. It is important to understand that the construction-specific PPPs so derived should only be used when measuring and comparing volumes of construction output; they are not exchange rates and they should not be used when construction in one country is to be paid for using funds from another (e.g. where a client in the United States wishes to build a facility in another country and will be funding the project using US money). In such situations the use of current exchange rates to express costs in the same currency is appropriate. If the US client is choosing between potential locations in several other countries, then bringing all costs to a common base (most likely the client's national currency if the money to build will be sourced from the client's own country) is perfectly logical. If, however, the purpose is to answer the theoretical question – 'Is it more expensive to build in Country A, B or C?' – then the use of construction-specific PPPs is appropriate as the question then relates to the volume or amount of construction that a client can buy for a given amount of local currency.

Construction PPPs (CPPPs) are derived using cost data associated only with items relevant to the construction industry, usually a mix of materials, labour and equipment, weighted to reflect a typical mix of the cost of resources or inputs. Such a mix might be 55% materials, 40% labour and 5% equipment, but this varies according to location and the type of construction. Where labour is cheap and abundant, and many materials are imported, the cost of the labour and equipment components may shrink while materials costs rise, although percentages may not vary much because the prices and quantities of labour in different countries tend to cancel each other out (i.e. lots of cheap but relatively inefficient workers in poorer countries may make up much the same proportion of total costs as a smaller number of expensive but highly skilled workers in richer countries).

Where labour is more expensive, most materials are largely locally produced and more capital is invested in equipment, the mix will vary to reflect these differences. Similarly, the construction of civil engineering projects, such as highways and dams, will mostly use large quantities of basic materials (e.g. steel and concrete) while utilising more plant and machinery than is usual in the construction of buildings.

Temporal indices are also about volumes – they normalise, or bring to a common base, values over time (they deflate or convert current to constant prices) and allow volume comparisons to be made. Both temporal and spatial price indices convert values in nominal or current prices (prices of the day) to real or constant prices.

The diversity of output

Construction activity is typically divided into three broad categories: residential, non-residential and engineering construction; high-rise residential, due to its scale, may be included with non-residential. Within these categories there is great diversity in the projects (products) that are built. Few buildings are truly identical. Even when standard designs are repeated there are variations due to differences in local regulations, site, climate, availability of materials, influence of adjacent structures and more. Engineering construction is perhaps even more diverse. It covers a disparate set of project types ranging from dams and power generation plants to railway lines, tunnels and bridges and much more.

Within this diversity, when looking to make comparisons between countries (say) it is necessary to find items (whether products, materials, components or whole projects) that are similar enough to be compared in a rational way yet are sufficiently representative of 'typical' construction in the countries being compared. This lies at the very heart of any method for comparing, for example, projects across countries.

There are other divisions of total construction output; for example, the construction of new buildings, improvement of existing buildings (i.e. additions, major renovation and refurbishment) and routine R&M of existing buildings. They involve construction activity, but each has its own characteristics and how such activity is recorded varies. A good deal of R&M is done by owners as do-it-yourself (DIY) projects and may well never be recorded beyond the purchase of materials and components.

The complexity of output

Any sort of measurement of the construction industry, particularly where some sort of comparative exercise is being undertaken, tends to rely on data that is for 'average' conditions or 'standard' or 'typical' projects. Buildings are, however, seldom standard products and they vary in many ways; apart from the more obvious differences in type/function, scale/size and location, differences in the complexity of buildings affect both the cost and the time required to build them. Such differences are a key concern in comparative studies.

Tilley *et al.* (1997) created a project performance index based on the number of requests for information (RFIs) generated during construction. As it would be expected that the construction of larger and/or more complex projects might lead to a greater number of RFIs, some sort of correction was required to account for differences in scale and complexity. The performance index they produced is:

$$PI_1 = \frac{N_c}{CV \times D}$$

Where: N_c = number of *information clarification* type RFIs
CV = estimated final contract value ($100,000's)
D = initial project duration (months)

The factor $CV \times D$, the product of construction cost and time, was used as a measure of complexity to offset the expected increase in incidence of RFIs in larger / more complex projects. Langston and Best (2001) used a similar factor in their performance index for projects. More recently, Langston (2015) investigated that index further and then derived a construction complexity index based on time, cost and floor area that could be reduced to the ratio of the cost squared to the floor area squared.

Statutory/regulatory requirements

Building codes and other regulations (e.g. workplace health and safety legislation) vary considerably from country to country, and these naturally have an effect on the design and construction of buildings. These varying requirements can affect both time and cost. Such provisions may result in physical differences between buildings or through different mandated work practices.

Fire safety requirements relating to building materials are an example of the way that physical differences in buildings are created by differing codes and regulations. Façades on high-rise buildings have been the subject of much scrutiny since the tragedy of London's Grenfell Tower, with the belief that the fire spread quickly due to the nature of the external cladding (Davey 2017). In Australia, AS5113 sets strict standards in regard to the assessment and classification of external walls of buildings according to their tendency to spread fire. These standards are more demanding than those current in other parts of the world, even in developed countries such as the UK and US, and these stricter controls can have a significant impact on building costs (Ervine 2018).

Site safety requirements vary from place to place, as does the degree of enforcement of such regulations. In Australia, for example, all electrical tools have to be checked and tagged every three months by a licenced technician; sub-contractors are required to supply copies of the certificate of currency for their workers' compensation insurance every month; and most sites with 20 or more people on-site will have a full-time safety officer. All of these requirements add to the cost of construction through increased administration costs and/or additional salaries. These are just a few of the quite stringent safety regulations in Australia that add to building costs there, but which do not necessarily apply in other places.

The quality of output

Differences in the quality of buildings add to the difficulty of measurement of output, particularly when the aim is to compare industry characteristics such as productivity or efficiency between producers, whether at the project, firm, regional or national level. The quality of construction output can be considered in a number of ways including assessment of compliance with codes and regulations, surveying standards of workmanship, or assessment against recognised standards such as star ratings for hotels or office space which will reflect attributes such as the quality of materials and finishes specified for different projects.

While compliance might be taken as a given, codes can vary considerably between different jurisdictions and these differences can affect cost and productivity. Similarly, the diligence with which compliance is assessed and enforced varies from place to place. Informal construction and DIY building work typically will not be subject to any sort of compliance assessment.

In the past, assessment of workmanship was largely a subjective exercise; Flanagan *et al.* (1986: 4) suggested there was 'no recognized method of quality assessment' for buildings. McKim *et al.* (2000) identified several more objective measures including estimated cost of rework and repairs, and number of requests for rework and repairs. Brown and Adams (2000) considered 'delivered quality' and assessed that as a function of the number of building defects recorded at the handover of completed buildings.

Sodangi *et al.* (2010) surveyed Malaysian construction clients to obtain their views on contractors' quality performance based on the clients' assessment of the buildings that contractors had completed for them. The parameters addressed included building performance, building reliability, compliance with design standards and specification, durability of buildings, serviceability and aesthetics. As the survey sought client opinions, it must be concluded that this sort of measurement is largely subjective. Furthermore, it is hard to see how a client's opinions on the aesthetics of a completed building somehow reflect the quality performance of the contractor. What this study highlights more than anything is how difficult it is to measure construction quality objectively.

Singapore, however, introduced the Construction Quality Assessment System (CONQAS) in 1989 (BCA n.d.), and this has been used and refined over a long period with the ninth edition published in 2017 (BCA 2017). Standards of workmanship are assessed against benchmarks with a view to reducing the degree of subjectivity involved. Variations of the CONQAS approach are used in a number of other countries, including China and Malaysia, but there is certainly no internationally agreed standard for measuring construction quality in any of its varied manifestations.

Conclusion

In this chapter, just some of the factors that complicate the measurement of construction have been reviewed. What is clear is that while measurement is often subject to many complications, there are serious efforts being made to find methods that address issues such as variations in quality and complexity that will eventually lead to better measurement and thus to more robust comparisons. Understanding a problem is usually an important first step towards solving that problem and while some of the chapters that follow do offer some possible paths towards better understanding through new methodologies, often the value lies as much in the exploration and analysis of the problems as much as in any potential solution that is put forward.

There are more topics that have not been addressed in this book or its companion volume; these include further investigation of the value of informal construction in developing countries, the performance of construction firms (performance in terms of economic performance rather than the more common measures such as timely completion of projects and completion within budget) and, as indicated earlier, comparative studies of what is included in or excluded from national construction statistics. And more work is needed on some, if not all, of the topics that have been addressed.

Note

1 '[I]t should be obvious that in countries where food production is not based on wheat, sesame seeds, beef, dairy products, dill pickles and potatoes and where a *Big Mac* is a luxury item, available only in major cities to urban elites mimicking Western tastes, rather than a fast food staple, it is not any sort of a "standard commodity"' (Croce *et al.* 1999: 21).

References and further reading

BCA (n.d.) *Construction Quality Assessment System (CONQAS)* (Singapore: Building and Construction Authority). February 2018. www.bca.gov.sg/Professionals/IQUAS/conquas_abt.html.

BCA (2017) CONQAS ® *The BCA Construction Quality Assessment System* (Singapore: Building and Construction Authority). www.bca.gov.sg/Professionals/IQUAS/others/CONQUAS9.pdf

Best, R. and Meikle, J. (2015) *Measuring Construction: Prices, Output and Productivity* (Abingdon: Routledge).

Brown, A. and Adams, J. (2000) Measuring the effect of project management on construction outputs: A new approach. *International Journal of Project Management*, **18** (5), 327–335.

Croce, N., Green, R., Mills, B. and Toner, P. (1999) *Constructing the Future: A Study of Major Building Construction in Australia*. Newcastle: Employment Studies Centre, University of Newcastle.

Davey, E. (2017) Grenfell tower: Polyethylene cladding on scores of towers. *BBC News.* 24 October. www.bbc.com/news/uk-41680157.

Deming, W. E. (1994) *The New Economics for Industry, Government, Education* (Cambridge, MA: The MIT Press).

Ervine, A. (2018) *Personal Communication.* Dr Adam Ervine, fire engineer, WSP Consulting, Brisbane.

Eurostat-OECD (2012) *Eurostat-OECD Methodological Manual on Purchasing Power Parities* (Luxembourg: Publications Office of the EU).

Flanagan, R., Norman, G., Ireland, V. and Ormerod, R. (1986) *A Fresh Look at the UK & US Building Industries* (London: Building Employers Confederation).

Habitat International (1978) Duccio Turin: Bibliography. *Habitat International*, **3** (1–2), 19–29. https://doi.org/10.1016/0197-3975(78)90030-9.

Harrington, H. J. (1987) *The Improvement Process* (New York: McGraw-Hill).

ICP (2015) *International Comparison Program*, World Bank. http://go.worldbank.org/PQ5ZPPYSY0.

Langston, C. (2015) Performance measures for construction. In: Best, R. and Meikle, J. (eds.) *Measuring Construction: Prices, Output and Productivity* (Abingdon: Routledge).

Langston, C. and Best, R. (2001) An investigation into the construction performance of high-rise commercial office buildings worldwide based on productivity and resource consumption. *International Journal of Construction Management*, **1** (1), 57–76.

Lethbridge, A. (2008) *Regency Ramble*, 28 August. http://regencyramble.blogspot.com.au/2008/08/.

McKim, R., Hegazy, T. and Attalla, M. (2000) Project performance control in reconstruction projects. *Journal of Construction Engineering and Management*, March/April, 137–141.

Meikle, J. and Grilli, M. (1999) Measuring European construction output: Problems and possible solutions. In: Ofori, G. (ed.) *Proceedings of Construction Industry Development in the New Millennium*, 27–29 October (Singapore). https://pdfs.semanticscholar.org/c642/e792e413d639693f41838f4ed63d0c7c987f.pdf.

Meikle, J. and Thomas, P. (2012) *Calculating Construction PPPs*. International Comparison Program, 7th Technical Advisory Group Meeting (Washington, DC: World Bank). http://pubdocs.worldbank.org/en/848861487262353092/03-04-ICP-TAG07-DRAFT-CalculatingContructionPPPs.pdf.

Rawlinsons (2017) *Australian Construction Handbook*, 35th ed. (Perth: Rawlhouse Publishing).

Sodangi, M., Idrus, A. and Khamidi, M. F. (2010) *Measuring Quality Performance in Construction*. International Conference on Sustainable Building and Infrastructure (ICSBI 2010), 15–17 June 2010. Kuala Lumpur Convention Centre. www.researchgate.net/publication/267232541_Measuring_Quality_Performance_in_Construction.

Tilley, P., Wyatt, A. and Mohamed, S. (1997) Indicators of design and documentation deficiency. In: Tucker, S. (ed.) *5th Annual Conference of the International Group for Lean Construction* (Australia: Gold Coast), 16–17 July, 137–148. www.iglc.0net/Papers/Details/31.

Editorial comment

In the preceding chapter the question was asked: how do we delimit the construction industry and thus define what constitutes the industry? Where we place the boundaries of the industry that we measure will naturally affect the results of our measurement.

At the beginning of the 21st century, the Organisation for Economic Co-operation and Development (OECD 2001: 66) stated that:

> The contribution of the construction industry to GDP in OECD member countries ranges between 5%–8% and between 5%–9% of total employment. Of course, given that a significant proportion of construction activity is undertaken by units *outside the construction industry* [emphasis added], these percentages understate the importance of the construction industry.

There is some irony here: if units 'outside the . . . industry' are carrying out construction activities, then why are they not considered to be part of the industry? The answer may lie in the way that construction activity is measured in national accounts, which is usually by measuring the value of such activity or the value added through such activity. Such measurement is typically focused on value added through on-site building activity only, and this limited view of the industry typically does not take into account a great deal of associated activity that is not directly part of on-site construction. The measurement is further complicated by factors such as the varying degrees of off-site fabrication that are employed in different projects and how the value of that component of projects is captured in the measurement.

The existence and importance of the wider industry is recognized, and the two views of the industry have been called 'broad' and 'narrow', with the narrow industry defined as on-site work and the broad industry as one that also includes the supply chain of materials, products and assemblies as well as professional services. An example of the complexity of this concept is provided by the emergence of large multi-disciplinary firms that not only manage construction but also contribute to both architectural and engineering design (for example); this has led to significant blurring of the lines of demarcation between those providing professional services (part of the broad industry) and those managing and/or carrying out on-site construction (the narrow).

In this chapter, author Gerard de Valence looks at the importance of measuring the broad industry as well as the narrow and shows that this could be done by producing satellite accounts for the broad industry. He proposes the recognition of a 'built environment sector' that represents the broad industry and suggests that satellite accounts for such a sector could be created using data that is already available to many national statistical offices.

2 Accounting for the built environment

Gerard de Valence

Introduction

The way we see and understand an industry typically starts with the economic data we get from the national accounts and other collections done by national statistical agencies. This data is often presented in tables and graphs showing industry shares of the economy or gross domestic product (GDP), so the whole economy is seen as a number of distinct groups of industries, such as manufacturing, retailing or business services. Over the decades since the 1930s, when economists in America and Britain began developing the concepts, methods and processes required to measure the economy, a complex and sophisticated system has evolved (Vanoli 2005). However, as with all extensive data collections, issues around the definitions and categories used to order the data are ever present, and the *System of National Accounts* (SNA), first published by the United Nations in 1953, is now in its fourth edition (UN SNA 2008; previous revisions were published in 1968 and 1993).

Although the SNA is used internationally for measuring economic activity, countries adapt it so it better suits the structure and characteristics of their economies, the industries present and their data collection methods and systems. Over time, new industries have been added to the total included in the SNA to reflect the increasing complexity of modern economies, with the creation of new categories and revisions of the definitions used to establish the boundaries of industries. In some cases an industry is divided into parts, for example separating Retailing and Wholesaling into two industries; in others the changes happen through revised definitions of the forms of economic activity included or excluded from an industry – a particular issue with industries when they are growing rapidly, as finance and telecommunications did after 1980, for example. The definition of construction in the SNA, as the value of on-site work done by contractors and sub-contractors, has not, however, been changed.

Output and employment are the two most important pieces of economic data. For construction, the national accounts data on industry output excludes two important components of building and construction on-site work. First, it does not include work done in-house by organizations such as retail chains and public agencies that are not classified as construction contractors but undertake major

Table 2.1 Building and construction industry statistics (*Standard Industrial Classification of All Economic Activities*, UN Statistical Division 2008)

Sector	Type
Residential building	Detached housing, medium and high density dwellings, includes alterations and additions.
Non-residential building	Private – retail, commercial, industrial, hotels, etc. Public and social – education, health, community, etc.
Engineering construction	Bridges, ports, rail, mines, electricity, roads, water and sewerage, dams, telecommunications, etc.

projects and have large-scale maintenance and refurbishment programs. Given the size of many of these organizations, the cumulative value of this excluded work is significant. Second is work done by households and other forms of informal building, which is work done by owner builders, do-it-yourself (DIY), cooperatives, communes and the like. This activity is not included in industry statistics but is loosely linked to the rest of the industry through sales of equipment, materials and components and so on. The effect of these omissions is a large, and largely unknown, contribution to the economy from construction that is attributed to other industries in the SNA.

In the same way as the SNA provides a system to measure the economy, there is the *International System of Industrial Classification* to measure industries. This is also published by the UN Statistical Division (2008), and again many national statistical agencies produce their own versions of its standard industrial classification (SIC). For building and construction, statistics are collected by sector and then divided into building or structure type, shown in a generalized form in Table 2.1. Projects within a defined sector are then grouped together to establish sector size and importance – detached housing for example, or commercial developments. Because the data on industry activity and output is presented in these classifications, analysis of trends and forecasts of building and construction work are also usually found in this format.

The next two sections look at different definitions of the construction industry, and research around its scope and boundaries within the limits of the SIC. This is followed by a proposal for using industry data to develop a measure that includes as much as possible of the economic activity associated with the production and maintenance of the built environment.

Narrow and broad definitions of construction

The SIC definition of the construction industry captures the on-site activities of contractors and sub-contractors, and this data on building and construction work is taken to represent the industry. However, the on-site work links suppliers of materials, machinery and equipment, products and components, and all the other inputs required to deliver the buildings and structures that make up the built environment. Consultants provide professional services such as design, engineering, cost planning and project management services as inputs into building and construction. There are also inputs from urban planning, transport, finance and legal services. This can be thought of as the difference between the broad construction industry, made up of contractors and sub-contractors supported by equipment suppliers, professional services, materials, manufacturers, distributors and others, and the on-site work that is measured as 'construction activity', which only includes contractors' purchases of materials and supplies for use on building sites, and does not include consultancy services or construction work by organizations other than contractors.

Following Ive and Gruneberg (2000), Pearce (2003) called these the broad and narrow industry structures. The narrow industry was defined as the on-site work reported as industry output in the national accounts, and the wider industry as 'the supply chain for construction materials, products and assemblies, and professional services such as management, architecture, engineering design and surveying' (Pearce 2003: 10). Where to draw the boundaries of the wider industry is an open question, as a diverse range of firms, professional institutions, government regulators and authorities all contribute to the creation and maintenance of the built environment. Turin (1969) and the Bartlett International Summer School series in the 1980s (Groak 1990) advocated looking at the sector that produces the built environment in broad and integrative terms, and a framework that captures the extent of that diversity is shown in Figure 2.1.

Research has focused on the definitional aspect of this wider construction industry. Pearce (2003) went through the SIC definitions of industry groups and classes in order to identify the backward and forward linkages that are relevant to the on-site work done by the construction industry, which is the narrow definition. Figure 2.2 shows the contributing industries and inputs Pearce includes in the broad industry. Similarly, Squicciarini and Asikainen (2011) provided a thorough comparison of the European and North American SIC classes that can be included in a wider definition and extended the scope of their 'wide industry' further than Pearce. They concluded that measures of composition, structure, value added, skills and research and development (R&D), and the output of the construction sector change substantially when a broader definition of the sector is used. In Squicciarini and Asikainen, the industries included are drawn from contributors to the three project stages they used to represent the industry, in what they call a value chain approach: preproduction activities and services; core production; and postproduction activities and services.

There have been attempts to estimate the economic contribution of the wider construction industry. Using what they called a 'meso-economic' approach (i.e. between micro and macroeconomics), Carassus et al. (2006) compared the size of the 'construction sector system' in seven countries, again using a very wide definition to include property management, repair and maintenance and the institutional actors involved; because the necessary data is not available for all the activities included, these estimates are somewhat speculative for some countries. Ruddock and Ruddock (2009) also estimated the size of the construction sector

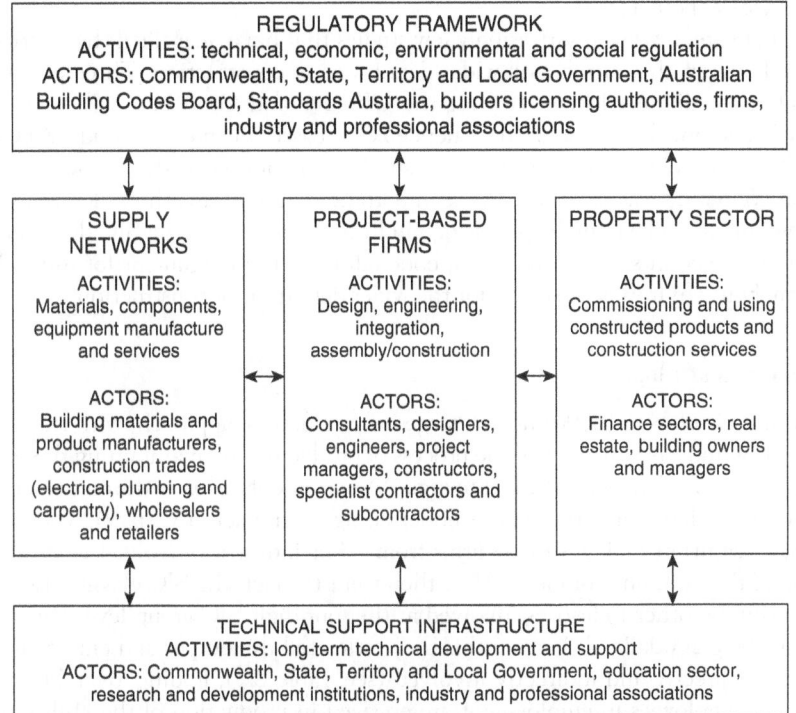

Figure 2.1 A framework for the built environment (Gann and Salter 2000)

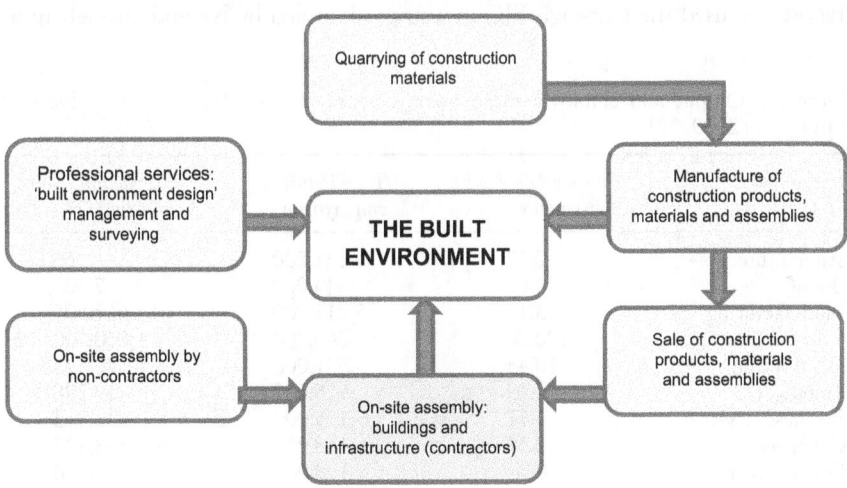

Figure 2.2 Broad and narrow industry structures (Pearce 2003)

for 20 European countries, which ranged between 12% and 22% of GDP with an average of about 17%.

There are, however, surprisingly few studies that have quantified the relationship between the narrow and wider definitions of construction. Three previous studies are discussed here: two for the UK (Ive and Gruneberg 2000; Pearce 2003) and one for Australia (de Valence 2001, based on an earlier study; AEGIS 1999). The relevant data from these studies is reviewed in the following section, followed by a discussion on the importance of measuring the wider industry. How this could be done by preparing satellite accounts from the SIC data in the national accounts is also discussed, concluding with an argument for using the term 'built environment sector' for the wider definition of construction.

Previous studies

Ive and Gruneberg (2000) used a 'stages of production' approach, with the construction industry one part of the process of producing and maintaining the built environment. They were the first to argue 'A *narrow* definition of the construction industry includes only those firms undertaking on-site activity' (p. 9), whereas a *broad* definition includes many firms from other industries involved in production of the built environment. They then went through the SIC classification to identify the other industries involved at the four-digit SIC group level (Ive and Gruneberg 2000: 25–28). From this they estimated the output and employment of firms directly and indirectly involved (see Table 2.2), finding 'over 11.2 per cent of employees in employment are engaged in production of the built environment. This compares with the figure of 4.7 per cent taking the construction category alone' (Ive and Gruneberg 2000: 11).

This research was the basis of the industry estimates given in Pearce (2003), which adopted the narrow and wide terminology of Ive and Gruneberg. The Pearce report also used the four-digit SIC industries identified by Ive and Gruneberg to

Table 2.2 Output and employment in broad construction in the UK, 1993 (Ive and Gruneberg 2000: 12)

Industry	Output % going to construction	Total industry employment	Construction related labour share of output
Agriculture	0.02	281,000	66
Energy	1.65	443,000	7,302
Manufacturing	13.12	5,012,000	657,805
Construction	100.00	1,060,000	1,060,000
Distribution	10.43	3,551,000	370,372
Transport	2.09	925,000	19,298
Business services	6.77	3,135,000	212,312
Other services	2.50	6,748,000	168,437
Other sectors	0.0	1,198,000	0
Total		22,353,000	2,495,592

quantify the relationship between the narrow and broad definitions of the con-struction industry by including the contribution of related industries. Pearce found, for the UK in 2001, using the number of firms and the gross value added in five industry groups, in both cases contractors accounted for around half the total, as shown in Table 2.3. Gross value added (GVA) is calculated as the difference between the costs of the inputs used and total industry output, and when GVA for all the industries that make up an economy are added together, they give GDP measured by the production approach (UN SNA 2008).

The Statistical Appendix to the Pearce Report (see Table 16 in Pearce 2003) then took three- and four-digit industry SIC data and, broadly, got the same result for output and employment in five sectors. Table 2.4 shows this data for the 26 three- and four-digit classes and groups identified as relevant, aggregated into their parent SIC industry groups. Ruddock and Ruddock (2009) also took this approach to SIC data and built on it to get their estimates for the construc-tion sector in 20 European countries. The dataset in Pearce is unusually detailed; however, Ive and Gruneberg and Pearce used the 2003 UK SIC, which has since been revised as the UK 2007 SIC. Table 2.4 also shows the percentage shares of construction in the totals for turnover, GVA and employment as 57, 55 and 57, respectively.

Table 2.3 UK shares of value added 2001 (Pearce 2003: 13–15)

Construction and related Industries	No. of firms	Share of GVA
Contractors	192,404	52%
Mining and quarrying of construction materials	2,248	2%
Manufacture of construction products	20,863	15%
Sale of construction products	81,997	15%
Professional services	57,636	16%

Table 2.4 UK construction and related industries 2001 (Pearce 2003)

SIC	Industry	Turnover/ % shares of total output		GVA at basic prices/ % shares of total output		Employment costs/ % shares of total output		Employment during year/ % shares of total output	
F	Construction	131,179	57	47,647	55	23,798	52	1,370,000	57
C	Quarrying	4,439	2	1,642	2	695	2	29,000	1
D	Manufacturing	40,948	18	15,265	18	9,044	20	465,000	19
G	Trade	24,228	10	5,135	6	3,019	7	155,000	6
K	Real estate, renting and business activities	30,207	13	16,814	19	9,247	20	399,000	17
	Total	231,001	100	86,503	100	45,803	100	2,418,000	100

In a similar analysis of the Australian industry, de Valence (2001) compared the size and scope of construction, first using data from the construction industry survey done by the Australian Bureau of Statistics (ABS 1998), and second from an industry cluster perspective. A cluster analysis includes all the firms involved in the construction process in an approach focused on the linkages and interdependencies between firms in a chain of production; this approach was applied to construction by the Australian Expert Group on Industry Studies (AEGIS 1999). AEGIS based their analysis on a division of the broad construction industry into five product-system segments: on-site services; client services; building and construction project firms; building products and supplies; and building fasteners, tools, machinery and equipment manufacturing. Each of these was then subdivided into product/service classes for collection of three- and four-digit SIC data from the Australian Bureau of Statistics (ABS 1998). Figure 2.3 shows the industries included in the framework for the construction industry cluster used as the basis of the data collection. Output and employment for these segments was found from the SIC industry sectors providing these products and services.

The AEGIS data for the broad construction industry in Australia in 1997 is shown in Table 2.5. When the data from both sources was compared, the size of the industry increased significantly, both in total income (and therefore share of GDP) and employment. The ABS industry survey income for 1996 was $58.6 billion, compared with total income of $110.4 billion from AEGIS, and total employment increased from 484,100 to 682,000.

When the industry segment shares are converted to percentages, with the *On-site services* and *Building and construction project firms* segments added together for

Building completion services	Installation trade services	Real estate services	Professional and technical services
On-site services		**Client services**	
Site preparation services	Building structure services	Residential services	Commercial services
HOUSE BUILDING	**OTHER RESIDENTIAL**	**NON-RESIDENTIAL**	**NON-BUILDING CONSTRUCTION**
Building products and supplies		Equipment hire and leasing	
Services and products		**Tools and equipment**	
Building materials	Structural building products	Building tools and fasteners	Construction machinery and equipment

Figure 2.3 AEGIS industry cluster (AEGIS 1999)

Table 2.5 Australian construction and related industries 1997 (AEGIS 1999)

Industry Segment	Total Industry Income $m	Total Industry Income %	Employment	Employment %
On-site services (trade services)	21,898	20	220,000	32
Building and construction project firms	34,250	31	108,000	16
Client services (engineering, technical, etc.)	8,607	8	102,000	15
Materials and product supplies	41,352	37	222,000	33
Machinery and equipment supplies	4,312	4	30,000	4
Total	110,419	100	682	100

the site-based measure of construction, the results look similar to the UK data above. In this case the narrow industry is 51% of income and 48% of employment of the broad industry.

Measuring the broad industry

While it would be preferable to have regular, detailed data on the size and scope of the broad construction industry, at present that is not available. Therefore, a heuristic, a general rule of thumb, is an alternative approach. Based on the studies above, it would be reasonable to conclude the broad construction industry might be twice the size of the narrow industry. Note this is not an accurate measure of the wider industry but a rough and ready estimate of its approximate scope and size. While this heuristic needs to be tested, it is likely to apply across many or most countries most of the time.

This heuristic is useful for four reasons:

1. It does not have to be exact to convey the importance of the wider construction industry to the economy;
2. It is based on regular and readily available data on (the narrow definition of) construction industry output, and is not as complex as the less frequent input-output data used to calculate multipliers and study industry linkages;
3. It naturally varies over time with fluctuations in both the business cycle and the building cycle, emphasizing the macroeconomic importance of the industry; and
4. It is not sensitive to local conditions, in that the share of the different industries included in the total will vary across countries, and the output shares of residential and non-residential building and engineering construction will vary over time, but the cumulative contribution of these industries to economic activity can still be estimated.

There is a way to turn this rough estimate into a more credible measure, and that is through the preparation of what is known as a satellite account, which reclassifies expenditures usually presented in different industry groupings into a single sector. These are used to provide more detail on sectors that are not adequately represented in the national accounts. The *System of National Accounts* published by the UN provides guidelines for national statistical agencies and, in Chapter 29, explains the reasons for preparation of satellite accounts and gives examples of their presentation. The process is similar to that used in Pearce (2003), where SIC data across industries is aggregated but done at a higher level of detail using the input-output or supply and use data from the national accounts.

At this time the most widely found satellite account is for tourism, so far produced at various times for nine countries, often jointly funded by industry and users. They have also been produced or proposed for a range of other industries such as health, the environment, R&D, information technology, infrastructure, non-profit institutions, human capital and households. While these are not normally produced annually and are sometimes not feasible at high levels of disaggregation, they allow re-use of existing data and thus maximize its usefulness. With the funding restrictions facing statistical agencies it is important to focus on the most important contributors to the construction industry. A version of satellite accounts known as *key sector accounts* selects a group of products or industries that are economically important and aggregates their data, and that could be tried as another approach.

The construction industry can be depicted in a variety of ways, but emphasizing how the built environment is created and maintained through project initiation, design, fabrication and construction to operation, repair and maintenance is most representative of the built environment sector as a whole. While the production of the built environment by the construction industry is an ongoing process, one we experience every day, it is just the most obvious part of a much larger system. The network of firms involved in maintenance of the built environment by the property management and real estate services industries, the manufacturers of fittings and finishings and plant and equipment, and the suppliers of materials and professional services is the broad construction industry, which brings together the participants in the production and management of the built environment. This extreme complexity in the number and range of activities involved in the built environment has, to date, prevented a coherent view developing of the industry and its linkages through the economy. In turn, this complexity has made efforts to improve the performance of the industry difficult and, as measured by productivity statistics, largely ineffectual (Vogl and Abdel-Wahab 2015).

Measurement is important, but so is choosing what to measure. In a time of rapid urbanization and great social and environmental challenges, the built environment and city policies have become central issues in public policy. The quality of the built environment the construction industry delivers is a major determinant of the quality of life. However, the data on cities, which policies are based on, and how cities are measured, typically follows the national accounting format of output and employment by industry. In this view construction is just one of many industries, but the irony is, of course, that it is the industry responsible for building cities. In a fundamental sense, how cities function depends on how well the building and construction industry can deliver the projects required.

How the building and construction industry organizes and delivers projects is constantly evolving, and there have been significant changes in the range of activities and types of firms involved in building and construction over the last two or more decades. The two trends underpinning these changes have been the increasing use of multi-disciplinary project teams (Gann and Salter 2000), where the boundaries between professional disciplines has become less distinct (Connaughton and Meikle 2013), and the 'make or buy' decision about provision has become more common. Facilities management is an example, an activity that used to be done in-house but is now often outsourced, sometimes but not always to construction contractors. Consultants bid for work as contractors, and contractors do consultancy (Bygballe *et al.* 2013). Urban planning was primarily associated with design but has become more extensively linked to real estate and development. This process of structural change in the industry is ongoing, as institutional roles and firm capabilities develop over time, and a satellite account would capture more of the economic contribution of the large number of diverse firms involved.

Meikle and Gruneberg (2015: 127) concluded in their analysis of international construction data that 'Governmental statistical agencies and industry commentators require improvements in the official definitions of construction and the way data is presented. There is a need for better information on the various measures of construction activity'. This refers to the narrow industry, but it is also true for the broad industry as defined above. At present the broad construction industry is not measured by any statistical agency, despite the importance of the built environment in economic development and social welfare. There is a need for better information on cities and the built environment, which a satellite account would provide. All else equal, better data means better policy and, by aligning the industry classifications, the cumulative economic contribution of the various industries in the built environment sector can be measured.

The built environment sector

Because building and construction is so diverse, it is hard to get an overview of the industry. With a vast variety of projects in all possible locations, made of materials ranging from primitive to rustic to ultra-sophisticated, the industry, particularly on a global scale (Runeson and de Valence 2008), is so broad that some system of classification and categorization is necessary. With the development of the Standard Industrial Classification, the industry has come to be defined by the data collected by national statistical agencies, but the role of the industry is much wider and deeper than the statistics show (de Valence and Lauge-Kirstensen 1998).

This means that the typical view of the industry is one called 'construction', and that industry is made up of three sectors: residential building, non-residential building and engineering construction. As well as the activity statistics for these sectors, there are other data sets. Many industry statistics use the division between building contractors, engineering contractors and sub-contractors (often called special trades). Because statistical data on work done and employment is presented in this form, most of the discussion and reporting of the industry also follows this pattern. This is not a bad thing but is not truly reflective of an industry as diverse and wide-ranging as building and construction.[1] Industry output statistics represent the industry as a set of functional projects, like detached housing or retail, and railways or hospitals, despite the fact that many buildings are mixed use. Also, there are many other ways of classifying and categorizing building and construction projects (e.g. Dubois and Gadde 2002; Shenhar and Dvir 2005).

Part of the problem here may be that the same term, 'construction', has been used in a number of different definitional and scoping studies. As the preceding discussion shows, it has a range of definitions and has been given widely different content in different studies. Thus Ive and Gruneberg (2000) had a 'construction sector', and Pearce (2003) described his approach as contrasting 'narrow' and 'broad' views of the industry. In Squicciarini and Asikainen (2011) the definitions are built around 'narrow' and 'wide' approaches and, like Ruddock and Ruddock (2009), they used the Pearce report approach with a 'construction sector'. Carassus (1998) had a 'construction system', and de Valence (2001) and Carassus *et al.* (2006) an industry 'cluster'. It would be helpful to agree on a common usage.

The term that arguably best encompasses the extraordinarily large number and range of participants in the creation and maintenance of the built environment, from suppliers to end users, is the *built environment sector* (BES). As indicated in Figure 2.4, there are many interconnected contributors and participants, with many regulatory, political and institutional factors involved within each of the eight categories used. On the left-hand side, there are products, design and construction, which make up the supply side of the BES. On the right-hand side, planning, property, building use and facilities management are the demand side of the BES.

Production and management of the built environment, how it is created through project initiation, design and construction and then repaired and maintained, requires a deep and dense network of firms known as a technological system (Hughes 1989). Together these firms make up the built environment sector.

Measuring the BES would help public policy and macroeconomic management for two reasons. First, the macroeconomic contribution of the BES to aggregate demand and employment is large, and possibly the largest in many countries. It is also one of the most volatile components of the economy, with annual rates of growth or contraction greater, and often much greater, than changes in GDP, making the BES a key driver of the business cycle. A satellite account collects those characteristics and thus provides data on trends in activity and output that have a significant effect on the national economy. Perhaps more importantly, changes in the composition of output of the BES would be a leading indicator of future demand as current new work completes, reflecting changes in the early stage project preparation activities required for future work. Through

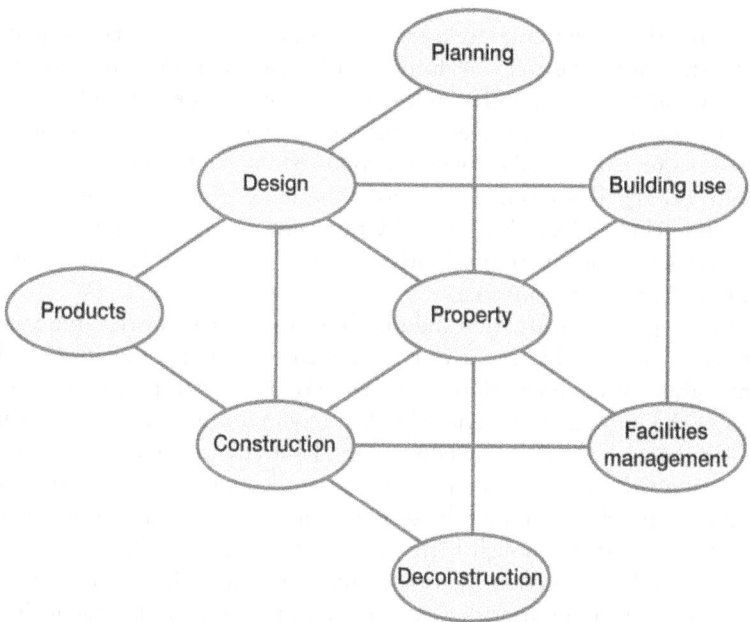

Figure 2.4 The built environment sector

industry linkages and lags, such slowdowns or pickups in project preparation can be strongly procyclical, amplifying the peaks and troughs of the business cycle. Because of the number of small firms found across the BES, the employment consequences of changes in activity levels are also significant.

Second, measuring the BES provides a way to measure the effectiveness of discretionary fiscal policy when that involves changes in expenditures on building and construction. Discretionary fiscal policy, as a response to the business cycle, is an increase in public spending to counteract a downturn in the business cycle or a recession, typically targeting public investment in both social and economic infrastructure. Tracking the impact of such expenditures through the economy is difficult but would show up in a BES satellite account. This would also allow a finer-grained analysis of the employment effects of different types of projects and programs.

In this context, it is worth noting that city policies involve significant infrastructure spending, and this is often their main focus. However, it is the associated induced development around the new infrastructure that drives longer-term growth. A satellite account would capture all that activity over time, thus giving a measure of the effectiveness of city policies in promoting urban growth and development. It may be that regional or city-scale satellite accounts would be most useful for urban planning and management.

Conclusion

The idea that the construction industry, as measured in the national accounts or using the standard industrial classification of industries, is only one part of the creation and maintenance of the built environment and the range of industries that encompasses is not new. Recognition of the industry's extensive linkages with other sectors, measured through the industry's high multiplier effects (Lean 2001; Gregori and Pietroforte 2015), gives the industry an important macroeconomic role (Lewis 2009; Meikle and Gruneberg 2015). Through those linkages the impact of construction activities on other parts of the economy is much greater than their direct contribution.

The building and construction industry links to other industries in a variety of ways, so measuring the extent of the built environment sector in a satellite account is the best representation of the dense network of firms involved in the creation of the built environment. The data required to measure the contribution to GDP and share of the economy of the built environment sector is available but scattered across separate industry data collections. This understates the importance of the BES and the role it can play in improving the performance of cities and urban spaces through better informed policy. A satellite account for the building and construction industry would also reflect changes in the range of activities and types of firms that contribute to the built environment. In the broad view of the industry, and in a satellite account, more of these activities would be included and the role of the sector better understood.

Note

1   The term 'building and construction' is in common use, although there is not necessarily any clear-cut explanation of the difference between the two component terms. It can be used to differentiate the construction of buildings (houses, schools, offices, etc.) from engineering construction (roads, bridges, airports, etc.) but the two terms are often used interchangeably and, just as often, one or the other is used alone to cover all types of building/construction, for example, the term 'construction industry' is very often used as a generic name for the broad industry that covers many types and aspects of construction; it may even include firms not directly engaged in the physical building process, such as suppliers and consultants. The International Comparison Program uses the term 'construction' to cover all types, and it separates construction activity into residential, non-residential and engineering construction, thus including 'engineering construction' as part of a generic category of activity called 'construction'. In this chapter the term 'building and construction' includes the range of activity covered by the ICP term 'construction'.

References and further reading

ABS (1998) *Construction Industry Survey 1996–97*. Cat. No. 8772.0 (Canberra: Australian Bureau of Statistics).

AEGIS (1999) *Mapping the Building and Construction Product System in Australia*. Australian Expert Group on Industry Studies (Canberra: Department of Industry, Science and Resources).

Bygballe, L. E., Håkansson, H. and Jahre, M. (2013) A critical discussion of models for conceptualizing the economic logic of construction. *Construction Management and Economics*, **31** (2), 104–118.

Carassus, J. (1998) Production and management in construction: An economic approach. In: *Les Cahiers du CSTB* (Paris: Centre Scientifique et Technique du Batiment, Livraison 395).

Carassus, J., Andersson, N., Kaklauskas, A., Lopes, J., Manseau, A., Ruddock, L. and de Valence, G. (2006) Moving from production to services: A built environment cluster framework. *International Journal of Strategic Property Management*, **10**, 169–184.

Connaughton, J. and Meikle, J. (2013) The changing nature of UK construction professional service firms. *Building Research and Information*, **41** (1), 95–109.

de Valence, G. (2001) Defining an industry: What is the size and scope of the Australian building and construction industry? *The Australian Journal of Construction Economics and Building*, **1** (1), 53–65.

de Valence, G. and Lauge-Kirstensen, R. (1998) Views on industry structure. *Form/Work*, **1** (2), 77–88.

Dubois, A. and Gadde, L. (2002) The construction industry as a loosely coupled system: Implications for productivity and innovation. *Construction Management and Economics*, **20**, 621–631.

Gann, D. and Salter, A. (2000) Innovation in project-based, service-enhanced firms: The construction of complex products and systems. *Research Policy*, **29**, 955–972.

Gregori, T. and Pietroforte, R. (2015) An input-output analysis of the construction sector in emerging markets. *Construction Management and Economics*, **33** (2), 134–145.

Groak, S. (1990) *The Idea of Building* (London: E. & F.N. Spon).

Hughes, T. P. (1989) *American Genesis: A Century of Invention and Technological Enthusiasm 1870–1970* (Chicago: University of Chicago Press).

Ive, G. J. and Gruneberg, S. L. (2000) *The Economics of the Modern Construction Sector* (London: Palgrave Macmillan).

Lean, S. C. (2001) Empirical tests to determine linkages between construction and other economic sectors in Singapore. *Construction Management and Economics*, **13**, 253–262.

Lewis, T. M. (2009) Quantifying the GDP construction relationship. In: Ruddock, L. (ed.) *Economics for the Modern Built Environment* (London: Routledge).

Meikle, J. and Gruneberg, S. (2015) Measuring and comparing construction internationally. In: Best, R. and Meikle, J. (eds.) *Measuring Construction: Prices, Output and Productivity* (Abingdon: Routledge).

OECD (2001) *Main Economic Indicators: Comparative Methodological Analysis, Volume 1: Industry, Retail and Construction Indicators* (Paris: OECD Statistics Directorate).

Pearce, D. (2003) *The Social and Economic Value of Construction: The Construction Industry's Contribution to Sustainable Development* (London: nCrisp).

Ruddock, L. and Ruddock, S. (2009) The scope of the construction sector: Determining its value. In: Ruddock, L. (ed.) *Economics for the Modern Built Environment* (London: Routledge).

Runeson, G. and de Valence, G. (2008) The new construction industry. In: Ruddock, L. (ed.) *Economics for the Modern Built Environment* (London: Routledge).

Shenhar, A. and Dvir, D. (2005) *Reinventing Project Management – The Diamond Approach to Successful Growth and Innovation* (Boston, MA: Harvard Business School Press).

Squicciarini, M. and Asikainen, A-L. (2011) A value chain statistical definition of construction and the performance of the sector. *Construction Management and Economics*, **29** (7), 671–693.

Turin, D. A. (1969) *The Construction Industry: Its Economic Significance and Its Role in Development* (New York: UNIDO).

UN Statistical Division (1968) *A System of National Accounts* (New York: Department of Economic and Social Affairs Statistical Office).

UN Statistical Division (2008) *International Standard Industrial Classification of All Economic Activities – Revision 4* (New York: United Nations). https://unstats.un.org/unsd/publication/seriesm/seriesm_4rev4e.pdf

Vanoli, A. (2005) *A History of National Accounting* (Amsterdam: IOS Press).

Vogl, B. and Abdel-Wahab, M. (2015) Measuring the construction industry's productivity performance: Critique of international productivity comparisons at industry level. *Journal of Construction Engineering and Management*, **141** (4), 04014085.

Editorial comment

'When I use a word', Humpty Dumpty said, in a rather scornful tone, 'it means just what I choose it to mean – neither more nor less'.

– Lewis Carroll

Terminology in construction statistics owes a lot to the Humpty Dumpty Principle, illustrated by the preceding quotation from *Through the Looking Glass*. In the first two chapters some variations in terms and definitions have been discussed, beginning with what the term 'construction industry' actually includes depending on the context in which it appears.

In this chapter, the author looks at some of the differing ways that construction is represented in national accounts and highlights the problems that this raises when attempts are made to compare data from national accounts from different countries. This examination demonstrates one aspect of the process of 'normalization' of data that is so often a necessity when making comparisons; the need to compare 'apples with apples' often means that some degree of adjustment of data is required. This ranges from simple conversions, such as the conversion from metric units to imperial units where data is gathered in one form in one place and a different form in another (e.g. plywood measured in square metres compared to square feet), to more problematic adjustments, such as substitution of locally available products for similar products that would not be in common use in a particular location (standard bricks, for example, may mean bricks of different sizes in different locations). At a broader level, in national accounts, the division of construction into categories such as residential and non-residential raises some basic questions; for example, which division should high-rise residential be included in? Residential construction ranges from small, single-family dwellings to apartment buildings with 100 or more floors (e.g. World One, under construction in Mumbai), with many types and sizes of building in between. If high-rise residential is included with high-rise commercial (non-residential), then where is the cut-off between residential and non-residential? And how do we account for fundamental differences between, say, a high-rise apartment building and an office building of similar size, where an apartment building will most likely include complete fitout of bathrooms, kitchens and the like and the office building is likely to be shell and core only?

Normalization of data often includes some degree of subjectivity and/or approximation and this may reduce the reliability of outcomes, yet there is often no option but to manipulate data if any sort of useful comparison is to be made. The analysis presented here outlines the broad problem of differences in categorization of construction activity in systems of national accounts and looks at some specific examples where the words used mean only what one person or agency intends them to mean – much in the manner of Humpty Dumpty.

3 Comparing construction in national industrial classification systems

Gerard de Valence

Introduction

The way we see and understand an industry typically starts with the data we get from the national accounts and other collections done by national statistics agencies. For building and construction, government statistics are typically collected by sector and then divided into building or structure type, shown in a generalized form in Table 3.1. Projects within a defined market are then grouped together to establish sector size and importance – detached housing for example, or commercial developments. Because the data on industry activity and output is presented in these classifications, analysis of trends and forecasts of construction work are also usually found in this format. Informal building is included here because it is an important part of the industry, but this sector is not usually included in industry statistics.

This chapter addresses those issues. Construction data is presented as part of a system of national accounts alongside the other industries that make up the economy. The next section outlines the system of national accounting used internationally, followed by an explanation of the system used to classify industries.

The chapter looks at the development and presentation of construction industry statistics reported by national and international statistical agencies. There is an *International Standard Industrial Classification* (UN 2008), but Europe, North America and Australasia have their own versions of the international standard. This raises the question of what the similarities and differences between their different versions of construction might be, and whether these differences are important.

The development of national accounting

National accounting is the collection and reporting of data for a whole economy. The first steps toward a national accounting system were taken in the 1940s based on the macroeconomic framework developed by Keynes (1936). The initial efforts of Clark (1933) in the UK and Kuznets (1937) in the US did not use an accounting framework, but all the elements for bringing together estimates of national income and output, consumer expenditure, government revenues and

32   *Gerard de Valence*

Table 3.1 Building and construction industry

Sector	Type
Residential building	Detached housing, medium and high density dwellings, alterations and additions
Non-residential building	Private – retail, commercial, industrial, hotels, etc. Public and social – education, health, community, etc.
Engineering construction	Bridges, ports, rail, electricity, roads, water and sewerage, dams, telecommunications, etc.
Informal building	Owner builders, do-it-yourself (DIY), cooperatives, communes, etc. Picked up in sales of equipment, materials and components.

expenditures, savings, capital formation, imports and exports and the balance of payments were present. By 1941, formalized national accounts were being prepared by the two governments based on the building blocks provided by researchers in national income and production (Stone and Stone 1977).

From the outset, national income, expenditure and output were the focus of policymakers, whose concern with stabilization policies stemmed from heightened expectations about the responsibilities and capabilities of government (Denison 1993). Sustained economic growth was the goal of economic policy and, as a consequence, the resources allocated to statistical collection reflected the macroeconomic concerns of government rather than the theoretical or empirical issues economists were pursuing (Triplett 1983).

The amount and range of the literature from the early days of development of the procedures and analytical techniques that underpin the national accounting framework is huge. Kendrick (1975) provides a comprehensive review by one of the founders of national accounting. Vanoli (2005) gives a history of the research effort during the years that national accounts and statistical agencies were being established and summarizes the contributions of pioneers such as Fabricant, Kuznets and Kendrick. Similarly, Stone (1988) and Ironmonger *et al.* (1988) discuss the development of national accounting in the UK and internationally, respectively. Stone (1988) details the struggle to get the early UK national income and expenditure accounts to balance – the major concern of economists at the time. Melvin (1995) gives a history of measurement issues in economics in a wide-ranging review with particular attention to issues that affect the services sector, as do Delavnay and Gadrey (1991).

Thus the modern form of economic accounts is a statistical system that has been under development for over 70 years. An accounting framework provides a useful picture of economic structure, defines the data required and simplifies the task of collection. When the estimates for income, expenditure and output (the triple identity) are derived from different sources, their accuracy can be checked during integration. However, the greatest difficulty statisticians face in measuring economic activity is its diversity (Stone and Stone 1977), and there are many possible sources of sample error in the surveys that provide the bulk of the data.

There are many reasons for data in the national accounts, its measures of quantities and values, to be inaccurate (see, for example, Stone *et al.* 1942), and the margin of error in national accounts varies from very low to very high depending on the type of data. Morgenstern critiqued national income and expenditure accounting because

> conceptual differences held among statisticians at different times and in different countries are bound to have decisive influence upon these statistics. Depending on the choice of one concept rather than another, the phenomena thus defined have their own error characteristics.
>
> Given these difficulties, it is easy to understand that conceptual changes are frequent. . . . One must also ask whether it is the constantly changing nature of the economy that calls for these conceptual revisions or whether they are an expression of our inability to settle conceptual issues.
>
> (Morgenstern 1963: 7)

The internationally accepted standard for presentation of national accounts is the United Nations System of National Accounts (SNA), first issued in 1963 and revised in 1968. It lays out guidelines for the methods to be used in data collection and analysis which are then interpreted and applied by individual countries, with some idiosyncratic results (Gordon 1990). Therefore, analysis based on data from any country's statistical bureau reflects the conventions followed in the collection and definition of that data in that country. These conventions differ between countries, so international comparisons have to be treated with circumspection, particularly when comparing construction output (Meikle and Gruneberg 2015). Also, data methodologies change over time within countries (Foss 1983), which further complicates the long-term analysis necessary in estimation of structural trends in the economy, such as the increase in the share of services in gross domestic product (GDP).

The national accounts provide a country's economic identity, with their measures of income, expenditure and output, and GDP and GDP per capita. However, these aggregate measures are built up from the data gathered at the level of households and firms, through their expenditures on products and services. While measuring the output of products is relatively straightforward, measuring the output of services ranging from personal trainers and hairdressers to neurosurgeons and accountants is more difficult. The units of output of economic activity are neither standardized nor homogeneous, and this is particularly the case for services (Denison 1962). Further, dynamic economies are characterized by frequent improvements in the quality of many different services and products as new ones are invented and introduced, old models are discontinued and more products are custom built. The building and construction industry is an archetype of an industry with non-standardized products; other such industries are shipbuilding and aircraft manufacture, and many professional services are similarly not standardized.

Non-standard products are a problem for output measurement. If custom goods account for only a relatively small proportion of output, the price indices used in the standardized part of the industry can be applied. Where most or all the output of an industry is customized, various cost indices are compiled, but such indices are inadequate deflators because they do not allow for changing overhead and profit margins or for changes in productivity and/or quality (Kendrick 1975). Price indices do not reflect quality improvements to services and products from technological developments, and this has been a major measurement problem for a long time. Measurement of the emerging digital economy is presenting a new set of challenges to statisticians.

Reprising Cannon (1994) on the general inadequacy of many industry statistics, Briscoe (2006) argued that in the UK, which has historically had one of the better national statistical agencies, there were serious problems:

> Problems with reliable and accurate data collection and statistical analysis include defining the scope and coverage of the industry; measuring industry outputs and their allocation across different types of activity; identifying construction firms; and measuring capital formation and capital stock, inconsistencies in employment statistics and labour market variables, discrepancies in measuring productivity, and the lack of international comparison. Needed improvements include a wider definition of construction output and employment activities, a continuous review of the usefulness and reliability of the published construction statistics, and the resolution of the discrepancies between different registers.
>
> (Briscoe 2006: 220)

The point about the scope of the industry is an important one, and is developed elsewhere in this book (see Chapter 2). As well as Briscoe, arguments for extending construction beyond on-site activities have been made in various ways by de Valence and Lauge-Kirstensen (1998), de Valence (2000), Pearce (2003), Carassus (2004), and Carassus et al. (2006), and more recently by Squicciarini and Asikainen (2011) and Meikle and Gruneberg (2015).

The Standard Industrial Classification

The main macroeconomic variables in the national accounts are consumption, investment, government expenditure and net exports. These aggregates are at too high a level to capture all economic activity, particularly the activity of companies (in the market sector) and other organizations (in the non-market sector, such as health or non-profit). An industry classification system collects companies and other organizations into groups with similar characteristics. The first *Standard Industrial Classification of All Economic* Activities (SIC) was established in the United States in 1937, with the United Nations *International Standard*

Industrial Classification (ISIC) following in 1948. This had its most recent revision in 2008:

> This fourth revision of ISIC enhances the relevance of the classification by better reflecting the current structure of the world economy, recognizing new industries that have emerged over the past 20 years and facilitating international comparison through increased comparability with existing regional classifications.
>
> (UN Statistical Division 2008: 3)

Since 1948 most countries have used the ISIC classification system or have developed national classifications derived from it, and while today there are many national variants on the ISIC format there is also a great deal of commonality. Economic activities are subdivided in a four-level structure. Activities are first divided into 'sections', which are alphabetically coded. These sections divide productive activities into broad groupings such as 'Agriculture, forestry and fishing' (A), 'Manufacturing' (C) and 'Information and communication' (J). The classification is then organized into numerically coded categories, which are two-digit divisions (see Table 3.2), three-digit groups, and four-digit classes (which have the greatest level of detail).

Table 3.2 ISIC categories

Section	Divisions	Industry
A	1–3	Agriculture, forestry and fishing
B	5–9	Mining and quarrying
C	10–33	Manufacturing
D	35	Electricity, gas, steam and air conditioning supply
E	36–39	Water supply; sewerage, waste management and remediation
F	41–43	Construction
G	45–47	Wholesale and retail trade; repair of motor vehicles and motorcycles
H	49–53	Transportation and storage
I	55–56	Accommodation and food service activities
J	58–63	Information and communication
K	64–66	Financial and insurance activities
L	68	Real estate activities
M	69–75	Professional, scientific and technical activities
N	77–82	Administrative and support service activities
O	84	Public administration and defense; compulsory social security
P	85	Education
Q	86–88	Human health and social work activities
R	90–93	Arts, entertainment and recreation
S	94–96	Other service activities
T	97–98	Activities of households as employers; undifferentiated activities of households for own use
U	99	Activities of extraterritorial organizations and bodies

ISIC codes are therefore four-digit numbers representing industries whose members are assigned on the basis of common characteristics in products, services, production processes and logistics systems. Meikle and Gruneberg (2015), in their analysis of international construction data, assessed it as a far from perfect system of classification and concluded:

> The current ISIC breakdown of construction activity is not particularly helpful to any user groups. It requires distinctions to be made among residential, non-residential and civil engineering work. It does not distinguish between construction investment (new work and improvements) or construction consumption (repair and maintenance) or among publicly sponsored, privately sponsored and mixed-funded work. Detailed breakdowns of construction activity could also address the different providers of construction output: construction contractors, the informal sector, households and so forth.
>
> (Meikle and Gruneberg 2015: 126)

Construction in the ISIC

In this section the 2008 UN International Standard of Industrial Classification of All Economic Activities (ISIC, 4th Revision) is compared with the 2006 *Australian New Zealand Standard Industrial Classification* (ANZSIC), the UK's *Standard Industrial Classification of Economic Activities* (ONS 2007) and the *North American Industry Classification System* (NAICS) used by the United States and Statistics Canada (2007). The UK SIC is the same as the NACE Rev. 2 (Eurostat 2008) system used in the rest of the European Union (EU).

Section F in ISIC includes the complete construction of buildings (division 41), the complete construction of civil engineering works (division 42) and specialized construction activities or special trades, if carried out only as a part of the construction process (division 43). Also included is the repair of buildings and engineering works. ISIC has a larger set of three- and four-digit groups and classes than the UK, Australian and North American SICs, being intended to cover economic activity in many widely different countries.

The UK SIC, ANZSIC and NAICS share the same basic structure, based on ISIC. Construction of buildings is separated by building types broadly into commercial and residential construction. The other subdivision shared is engineering or civil engineering. All nations have an SIC code for 'Other' and/or 'Other specialized trade contractors or building activities and services'. However, Canada, the US, UK, Australia and New Zealand do not consider many of ISIC's subdivisions as applicable to their own construction sectors. There are also additional classes in their trade contractor groups (see Table 3.3 for examples).

The UK SIC 2007 refers to three-digit groups as 'activities'; ANZSIC refers to them as 'services'; and NAICS refers to them as 'contractors'. This forces one to question terminology, but what seems likely is that contracting, services and activities are meant to convey the part played by each trade or building activity

Table 3.3 Comparison of ISIC and ANZSIC

F	ISIC	E	ANZSIC
41	Construction of buildings		
410	Construction of buildings		
		3011	House construction
		3019	Other residential building construction
	4100 Construction of buildings	3020	Non-residential building construction
42	Civil engineering		
421	Construction of roads and railways		
		3101	Road and bridge construction
	4210 Construction of roads and railways	3109	Other heavy and civil engineering
		3299	Other construction services n.e.c.
422	Construction of utility projects		
	4220 Construction of utility projects	3109p	Other heavy and civil engineering
429	Construction of other civil engineering projects		
		3101	Road and bridge construction
	4290 Construction of other civil engineering projects	3109	Other heavy and civil engineering
		3211	Land development and subdivision

to a project or task. Terminology is not particularly helpful if comparisons do not use a standardized form. For example, the UN SIC code 4390 for 'Bricklaying and stone setting' corresponds to the ANZSIC 3222 'Bricklaying' and the NAICS 238140 'Specialty trade contractors: masonry contractors', but there is no matching code in the UK SIC classification.

The *Oxford English Dictionary* definition of bricklaying as a form of brickwork fits neatly into the traditional definition of masonry. Masonry and brickwork can mean the same activity. Therefore, we can assume that 'Masonry contractors' (NAICS) and 'Bricklaying contractors' (ANZSIC) correspond to more or less the same activity. This method of comparing systems is basically an assumption, as we have no evidence from either party on exactly what they mean. Because NAICS does not detail bricklaying, it must consider it part of the masonry subdivision.

Thus terminology is problematic when making comparisons between these classification systems. The general meaning of mason is split between bricklaying and masonry. Both are meant to define a trade and/or construction skill; in this case masons refer to the specific skill of stonework, mainly cutting and shaping stones, and may or may not include laying them, so the use of the words mason and masonry can apply to different activities. If, in the ISIC, ANZSIC, NAICS and UK SIC systems, masonry is only cutting and shaping stones – work which mostly happens off-site – does this mean it may not be captured as part of the construction industry? But if masonry is classified as manufacturing, would bricklaying remain part of the construction industry? Either masonry is an archaic term and does not belong as part of the construction industry anymore, or more plausibly bricklaying and masonry still account for broadly the same general economic activity. For the purposes of this study it is not unreasonable to assume that masonry and bricklaying refer to the same thing, and we can then reasonably compare the different systems of industrial classification (see Table 3.4).

The general conclusion from the tables is that all three are similar to ISIC, which provides the bones of their systems. The UK SIC is the closest, with the same SIC codes corresponding to almost the same economic activities. For example, in ISIC, Section F is 'Construction' and the first subdivision starts with the code 4100. In the UK SIC, section F is 'Construction' with the code 4110. At the three-digit level of classes there are many differences between the four systems, but these basically reflect variations in local conditions ('Siding contractors' in Canada) or industry practice ('Hire of construction machinery with operator' in Australia or 'Scaffold erection' in the UK).

There is an important point of difference at the higher two-digit level. While there are exceptions to the ISIC format, with every country subtly unique yet broadly similar, NACE and the UK SIC have one specific inclusion that makes it different to ISIC. The SIC group 41.1 'Development of building projects' (and its

Table 3.4 Three-country comparison with ISIC

United Nations ISIC	Australia and New Zealand SIC	UK SIC and NACE	Canada NAICS
Section F Construction	Section E Construction	Section F Construction	236 Construction of Buildings
4100 Construction of buildings	30 Building construction	41 Construction of buildings	
Construction of all types of residential buildings		41.1 Development of building projects	
Single-family houses	301 Residential building construction	41.202 Construction of domestic buildings	2361 Residential building construction
Multi-family buildings, including high-rise buildings			
Construction of all types of non-residential buildings	302 Non-residential building construction	41.201 Construction of commercial buildings	2362 Non-residential building construction
Buildings for industrial production			23622 Commercial and institutional building construction
42 Civil engineering	31 Heavy and civil engineering construction	42 Civil engineering	237 Heavy and civil engineering construction

class 41.10 of the same name) is part of the construction industry. Neither ISIC nor the other two systems compared here have this subdivision.

Subdivision 41.1

The greatest difference between the three national SIC codes is found in the NACE Rev. 2 (Eurostat 2008) and the UK SIC. The 2007 revision of UK SIC 2003 introduced changes at all levels of the industrial classification. The 2007 revision includes the unique subdivision, 41.1 'Development of Building Projects', which includes construction management and project management of construction projects. ISIC has no definition for this activity within the construction industry.

ANZSIC code 3211 'Land development and subdivision' includes excavation and site preparation but excludes business units whose 'legal subdivision of land without land preparation are included elsewhere in the classification system based on the primary activity of the unit'. ANZSIC also has subdivision 67 'Property operators and real estate services', not part of the construction division, which covers property rental or management.

Statistics Canada has code 2372, which like ANZSIC is labeled 'Land subdivision' and 'comprises establishments primarily engaged in servicing land and subdividing real property into lots, for subsequent sale to builders'. Also like ANZSIC, it excludes 'establishments that perform only the legal subdivision of land'. Nonetheless Canada's system is different from ANZSIC because it includes construction management or agency construction management. However, NAICS code 23721 excludes much of what is included in the UK's SIC code 41100. There, project management, 'a turnkey-type service involving the entire project, including feasibility studies, the arranging of financing, and the management of the contract bidding and selection process' is classified under 54133 'Engineering services' (NAICS 2018). NAICS is closer to the UK SIC than ANZSIC, but ANZSIC and NAICS are more similar than the UK's subdivision. However, the development of building projects is not given its own subdivision by the Australian Bureau of Statistics, therefore the activity is not being accounted for in the same way that it is in the UK and the European Union (EU).

Within the countries using NACE, like the UK, the contribution to the construction industry from the 41.1 category varies widely. Across the EU the reported activity statistics compiled by Eurostat for the 'Development of building projects' category vary from 20% of construction value added in Cyprus to nothing in Sweden. Figure 3.1 shows the share of 41.1 in total construction value added in 2015 and the average of that share for the eight years to 2015.

The inclusion of the subdivision 41.1 in NACE distorts the data on the European construction industry. It is unevenly and unreliably reported across the 28 countries, but for a few countries it is a significant share of construction output. This subdivision is not included in the output statistics for Australasia and North America, and thus affects international comparisons of the construction industry's contribution to the economy.

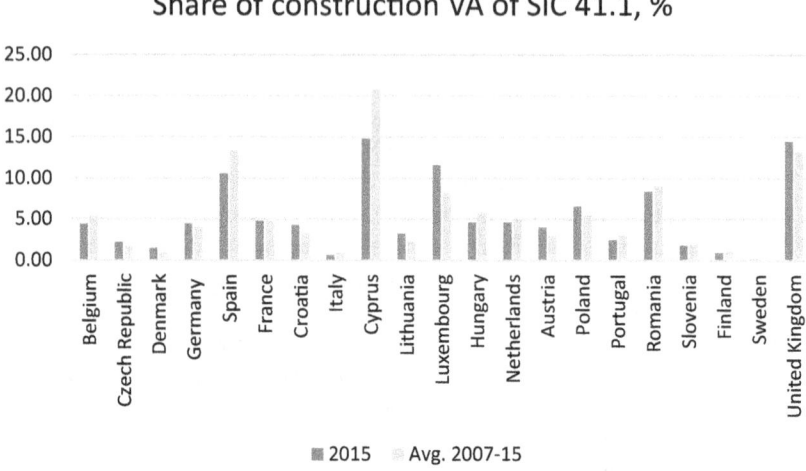

Figure 3.1 Development of building projects in EU countries (Eurostat 2017)

Conclusion

This chapter looked at the development and presentation of construction industry statistics as reported by national and international statistical agencies. Construction data is produced as part of a system of national accounts alongside the
other industries that make up the economy. With the system of national accounting used internationally, there is also a system used to classify industries.

Although there is an *International Standard Industrial Classification*, Europe,
North America and Australasia developed their own versions of the international standard. This raised the question of what the similarities and differences
between their different versions of construction might be, and whether these differences are important. In particular, do decisions on membership of construction
SIC industry classes and groups have implications for our understanding of the
role of the industry in international comparisons of GDP shares?

International comparisons are always fraught with difficulty. There are many
issues associated with exchange rates, purchasing power, traded and non-traded
goods and services, productivity and aggregation methods (Best and Meikle
2015). These issues apply as much, or more, at the industry level as they do to
national economies. The largest surveys of international prices have been done
by the World Bank's International Comparison Program (ICP), and the 2005
ICP found the average share of construction within individual country GDP as
measured by expenditure was 11.9%. There were major variations around this
average, ranging from a low of 1.6% in Nigeria's GDP to a high of 38.7% in
Bhutan's GDP (McCarthy 2013: 343), demonstrating how difficult international

comparisons are. Overall, construction's share of world GDP was 13%, adjusted for purchasing power parity (World Bank 2013).

The conclusion from the comparison of the ANZSIC, NAICS, NACE and UK SIC to ISIC, is that ISIC provides the bones of their systems. The UK SIC is the closest to ISIC, but at the three-digit level of classes there are many differences between the four systems. These basically reflect variations in local conditions through additional classes in their trade contractor groups. The NACE system in the UK and EU includes a unique subdivision, 41.1 'Development of building projects', which includes construction management and project management of construction projects. ISIC and the other systems do not include this activity in the construction industry. However, across the EU the share of construction output of 41.1 varies widely and is unevenly and unreliably reported across the 28 countries. For a few EU countries like the UK and Spain, this subdivision is a significant share of construction output, but because it is not included in the output statistics for Australasia and North America its inclusion affects comparisons of the construction industry both within the EU and internationally.

How construction industry data is collected and how the categories are defined are important; the former determines the quality and the latter the credibility of the statistics produced. This chapter has found the NACE inclusion of 'Development of building projects' in construction should be taken into account when making international comparisons, as the contribution of this subdivision 41.1 to industry output significantly increases the industry total in a few countries across the EU, but is either not reported or not included by countries using the ISIC, ANZSIC and NAICS systems.

References and further reading

ABS (2006) *Australian New Zealand Standard Industrial Classification* (Canberra: Australian Bureau of Statistics).

Best, R. and Meikle, J. (2015) The international comparison program and purchasing power parities for construction. In: *Measuring Construction: Prices, Output and Productivity* (Abingdon: Routledge).

Briscoe, G. (2006) How useful and reliable are construction statistics? *Building Research and Information*, **34** (3), 220–229.

Cannon, J. (1994) Lies and construction statistics. *Construction Management and Economics*, **12** (4), 307–312.

Carassus, J. (ed.) (2004) *The Construction Sector System Approach: An International Framework* (Amsterdam: CIB).

Carassus, J., Andersson, N., Kaklauskas, A., Lopes, J., Manseau, A., Ruddock, L. and de Valence, G. (2006) Moving from production to services: A built environment cluster framework. *International Journal of Strategic Property Management*, **10**, 169–184.

Carroll, L. (Charles L. Dogson) (1934) *Through the Looking-Glass*. Ch.6, 205. First published 1872.

Clark, C. (1933) The national income and the net output of industry. *Journal of the Royal Statistical Society*, **96** (4), 651–659.

Delavnay, J-C. and Gadrey, J. (1991) *Services in Economic Thought: Three Centuries of Debate* (Boston, MA: Kluwer Academic Publishers).

Denison, E. F. (1962) *The Sources of Economic Growth and the Alternatives Before Us* (New York: Committee for Economic Development).

Denison, E. F. (1993) The growth accounting tradition. In: Szirmai, A., Van Ark, B. and Pilat, D. (eds.) *Explaining Economic Growth: Essays in Honour of Angus Maddison* (Amsterdam, New York: North-Holland).

de Valence, G. (2000), reprinted (2010) Defining an industry: What is the size and scope of the Australian building and construction industry? *The Australian Journal of Construction Economics and Building*, **1** (1), 53–65.

de Valence, G. and Lauge-Kirstensen, R. (1998) Views on industry structure. *Form/Work*, **1** (2), 77–88.

Eurostat (2008) *Statistical Classification of Economic Activities in the European Communities* (Luxembourg: Office for Official Publications of the European Communities).

Eurostat (2017) *Construction of Buildings Statistics – NACE Rev. 2*. Statistics Explained. http://ec.europa.eu/eurostat/statistics-explained/index.php?title=Construction_of_buildings_statistics_-_NACE_Rev._2&oldid=327480.

Foss, M. F. (ed.) (1983) *The U.S. National Income and Product Accounts: Selected Topics*. NBER Studies in Income and Wealth, Vol. 47 (Chicago: National Bureau of Economic Research).

Gordon, R. J. (1990) *The Measurement of Durable Goods Prices* (Chicago: University of Chicago Press).

Ironmonger, D., Perkins, J. and Hoa, T. (1988) *National Income and Economic Progress: Essays in Honour of Colin Clark* (Basingstoke: Palgrave Macmillan).

Kendrick, J. W. (1975) *Economic Accounts and Their Uses* (New York: McGraw-Hill).

Keynes, J. M. (1936) *The General Theory of Employment, Interest and Money* (Basingstoke: Palgrave Macmillan).

Kuznets, S. (1937) *National Income and Capital Formation, 1919–1935* (New York: National Bureau of Economic Research).

McCarthy, P. (2013) Construction. In: *Measuring the Real Size of the World Economy: The Framework, Methodology and Results of the International Comparison Program – ICP* (Washington, DC: World Bank).

Meikle, J. and Gruneberg, S. (2015) Measuring and comparing construction internationally. In: Best, R. and Meikle, J. (eds.) *Measuring Construction: Prices, Output and Productivity* (Abingdon: Routledge).

Melvin, J. R. (1995) History and measurement in the services sector: A review. *Review of Income and Wealth*, **41** (4), 481–494.

Morgenstern, O. (1963) *National Income Statistics*, 2nd ed. (Harvard, MA: Harvard University Press).

NAICS (2018) *North American Industry Classification System (NAICS) Canada 2017 Version 1.0*, Statistics Canada. www23.statcan.gc.ca/imdb/p3VD.pl?Function=getVD&TVD=307532&CVD=307533&CPV=23&CST=01012017&CLV=1&MLV=5.

ONS (2007) *Standard Industrial Classification of Economic Activities* (London: Office of National Statistics).

ONS (2012) *Enterprise Count by UK SIC 2007* (London: Office of National Statistics).

Pearce, D. (2003) *The Social and Economic Value of Construction* (London: nCrisp).

Ruddock, L. (1997) Macroeconomic data related to the construction sector: An international overview. In: *First International Conference on Construction Industry Development* (Singapore: National University of Singapore), 232–239.

Squicciarini, M. and Asikainen, A-L. (2011) A value chain statistical definition of construction and the performance of the sector. *Construction Management and Economics*, **29** (7), 671–693.

Statistics Canada (2007) *North American Industry Classification System* (Toronto: Statistics Canada).

Stone, R. (1988) Progress in balancing the national accounts. In: Ironmonger, D., Perkins, J. and Hoa, T. (eds.) *National Income and Economic Progress* (London: Palgrave Macmillan).

Stone, R., Champernowne, D. G. and Meade, J. E. (1942) The precision of national income estimates. *Review of Economic Studies*, **9** (2), 111–125.

Stone, R. and Stone, G. (1977) *National Income and Expenditure*, 10th ed. (London: Bowes and Bowes).

Triplett, J. E. (1983) Concepts of quality in input and output price measures: A resolution of the user-value resource-cost debate. In Foss, M. F. (ed.) *The U.S. National Income and Product Accounts: Selected Topics*, NBER Studies in Income and Wealth, Vol. 47 (Chicago: National Bureau of Economic Research).

UN Statistical Division (1968) *A System of National Accounts* (New York: Dept. of Economic and Social Affairs, Statistical Office).

UN Statistical Division (2008) *Standard Industrial Classification of All Economic Activities* (New York: United Nations).

Vanoli, A. (2005) *A History of National Accounting* (Amsterdam: IOS Press).

World Bank (2013) *Measuring the Real Size of the World Economy: The Framework, Methodology and Results of the International Comparison Program – ICP* (Washington, DC: World Bank).

Editorial comment

In the opening three chapters the authors have discussed a range of challenges and difficulties that are part and parcel of measuring construction and/or accounting for construction in just some of its myriad forms. The discussion has ranged from measurement of individual on-site activities through to whole-of-industry measurements of construction activity, gross value added and so on.

Through his Brickonomics blog (https://twitter.com/brickonomics?lang=en), Brian Green has been an entertaining and intelligent commentator on the UK construction industry and its statistics for some years. In the light of ongoing criticism of official construction output statistics, this thoughtful chapter describes the evolution of UK construction output statistics since the Second World War and identifies some of the problems with the collection, analysis and presentation of the data.

Green's analysis highlights, in particular, the ever-changing environment in which the UK statistical exercises related to construction have been carried out, with numerous changes in both methodology and the people and agencies responsible for running them. It is little wonder that there have been significant changes in both how and what has been produced since the UK government, through the Ministry of Public Buildings and Works, began collecting construction statistics around 1940 as responsibility for construction statistics has passed between several government departments over the years before ending up with the Office of National Statistics (ONS) in 2008. During the 10 years that the ONS has looked after this sector there have been further changes in both methods and output.

The collection of construction statistics in the UK described here shows the variations that have occurred in one industrialised country; given the extent of the changes in the UK over time, it does not take much to imagine the variations that exist across both developed and developing countries. Many countries simply cannot afford the considerable cost of maintaining a large national statistical agency, and even in wealthier countries there is so much scope for the collection of statistical data that not all sectors of the economy can be serviced to the same extent. Construction must compete with other large sectors such as health, agriculture, and manufacturing and mining, and construction – as is the case with other industries – is not keen to bear the cost of compliance with government regulations that require firms to record, compile and supply data to national statistical offices.

For researchers and others, hoping to acquire construction industry data that is accurate, comprehensive and consistent over time is little more than wishful thinking, as the following chapter illustrates.

4   The challenges of measuring British construction output

Brian Green

Introduction

The construction industry has many measures of output. Various trade surveys seek to gauge activity across most aspects of contracting and building, the production and delivery of building materials and the construction-related professional and consultancy work.[1] As the hunger for economic data has risen, private sector data providers, such as Markit,[2] have joined what used to be mainly the preserve of trade associations.

Beyond these more direct indicators of activity, indices of estimated cost and price inflation (for example, see BCIS 2017) help in assessing the state of the construction market, and we can gain further insight into construction activity through measures of business investment (see ONS n.d.[a]) and household spending surveys (ONS n.d.[b]). Measures of the final products, such as homes built (Gov.uk n.d.), add further texture.

The construction output series for Great Britain (ONS n.d.[c]), which is currently produced by the Office for National Statistics (ONS), despite receiving much criticism of late, remains the benchmark for measuring the level of activity within the industry over time. The series has a long history stretching back to 1955, although there are estimates held by the Department for Business, Energy and Industrial Strategy (created in 2016) that extend the series back to 1932. Its data is used in calculating the national accounts and is used by industry forecasters as the baseline for forward projections of activity. The series is widely used and is currently the only measure that seeks directly to assess the actual volume of work undertaken within construction. Most other industry indicators tend to gauge rates of change.

It would, however, be unrealistic to expect such a statistical series over a period of 60 years to accurately reflect a consistent measure of construction activity. Responsibility for the collection and publication of the construction output statistics has shifted over time, and there have been significant changes in definitions, collection methods and types of work, and the structure and operation of the industry. Judgement calls, many long forgotten, will have been made over those 60 years even at the basic level of asking what actually constitutes the construction industry and what defines its output (ONS 2015a).

Responsibility for the construction output statistics can be traced back to the Ministry of Works which was formed in 1940 (Hansard 1940). Various construction statistics were collected by the ministry before the start of the current series. The earliest forms for this survey date from October 1954, when the Ministry of Works Statistics Division ran the enquiry known as the Quarterly Return of Employment and Output. The survey asked for the number of operatives employed by nine types of work and value of output for eight types of work.[3]

The Ministry went through various name changes and lasted until 1970 when it was absorbed into the Department of Environment (DoE). In 1997 the Departments of Environment and Transport merged to form the Department of Environment, Transport and the Regions. In 2001 responsibility for the construction statistics moved to the Department of Trade and Industry, which in 2007 became the Department for Business Enterprise and Regulatory Reform (BERR). The latest transfer of responsibility was in 2008, when ONS took over from the BERR, now the Department for Business, Energy and Industrial Strategy (BEIS).[4] By March 2009, the work previously carried out in BERR's Bristol and London offices moved to the ONS office in Newport, using the same systems and processes.

A project was set up to develop new methods and processes that would follow standard ONS methodology and systems. This led to the creation of a new construction output series, which was more in line with the approach taken by ONS to compile other short-term indicators, and that meant major changes were made which are reflected in the series from the start of 2010. Details of the development are covered in Crook and Sharp (2010).

The most obvious effect was to raise measured construction output by about 4%. At the request of users, back data from the previous series, despite some ONS reservations, was adjusted to fit to provide a continuous series. Importantly the periodicity moved from quarterly to monthly. So while the presentation of the data in the monthly tables might suggest a continuous series, there are significant differences in what was measured and how it was measured before 2010 compared to what is currently measured.

The timing of the transfer was, however, unfortunate. The construction industry had plunged into a deep recession and the UK government had brought forward investment into the sector to moderate the impact of falling private spending. With so much volatility and potentially conflicting signals, these were far from ideal circumstances in which to a replace a key indicator. Ideally a more stable period is preferred to allow for adjustments to various processes, such as seasonal adjustment.

There were concerns expressed by construction industry economists, analysts and users soon after the introduction of the new data series, not least because of the scale and frequency of revisions. Concerns were also raised by the Bank of England and HM Treasury. Meanwhile, a heightened interest in the economy, the spread of social media and an increased appetite in the media for statistics-based stories, led to increased interest in construction statistics, which in the past had tended to receive little attention from the media or anyone else outside the industry itself. This inevitably led to criticism from those who were poorly versed in the nuances of the industry and its data, who may not have appreciated the difficulties these present for statisticians. Consequently they may have had unrealistically high expectations regarding the accuracy with which construction can be measured. The credibility of the data in the eyes of the business media and analysts was then greatly undermined when the ONS had to withdraw a release on 12 August 2011, which contained a significant error (Kollewe 2011; Aldrick 2011).

Construction output statistics were the subject of a UK Statistics Authority inquiry later in the year following this major error. The statistical bulletin initially stated output growth initially for 2011 Q2 was 2.3%. That figure should have been 0.5%. The release commented that the 2.3% growth rate would have a consequential impact on gross domestic product (GDP), raising the estimate by 0.1%. This was at a time of intense interest in movement within the economy.

The statement put out by UK Statistics Authority also noted: "Construction statistics had been the subject of critical comment from analysts and journalists for some months. It had been suggested that in mid-2010 the series was overestimating output, and later in the year that the series was underestimating output" (UK Statistics Authority 2011a). In December 2014, the construction output statistics National Statistics designation was suspended (ONS 2014a).

The immediate reason for the suspension was a lack of confidence in the construction price and cost indices used to adjust data for the effects of inflation. Concerns over the production of these indices had been raised earlier in Dame Kate Barker's review of National Accounts commissioned by the ONS. The report stated: "In effect the source data for a number of the key price index series for construction activities have for a number of years come from a veritable 'black box'" (Barker and Ridgeway 2014). At the time, these price indices, which feed into the construction output series, were being produced by the BEIS, which had been working on improvements.

In April 2015 ONS took over responsibility and subsequently introduced an interim method for producing price indices while it sought to develop a long-term solution in-house.

ONS established a construction prices steering group to assist in the development of a solution for the price indices. This group became the construction statistics steering group, as it was decided to widen the scope and engage in a deeper review of the construction statistics.

This chapter looks at some of the challenges the ONS faces in seeking to produce robust statistics that measure construction activity and, importantly, a robust and meaningful measure of output that provides a consistent output data series to match the needs of its users and the reality of an industry in constant flux. It considers first the purpose of the construction output statistics and how they differ from other measures of changes in construction activity. It considers how construction is defined and the implications of the definitions and selection of firms which fall within the scope. It then looks at how the data is collected, how it is treated and how modelling is used to fill gaps for users. Finally, it explores the future prospects and challenges with particular attention to the impacts of changing technologies. It concludes with suggestions for how the statistics could be improved to deliver a more useful understanding of construction.

The purpose of the construction output statistics

The output statistics provide an essential estimate of construction contracting activity within the economy. They are the only regularly produced data that estimates the actual amount of activity undertaken by construction firms; most other industry surveys measure rates of change in various ways. The data is used within the UK national accounts in assessing the share of construction within GDP. The UK is also obliged to provide data to Eurostat on activity within industry, construction, retail and other services.

Beyond the statutory needs, the data is widely used by government and its agencies, including HM Treasury, BEIS, the Department for Communities and Local Government (DCLG) and the Bank of England. Construction output statistics are considered essential within the wider construction industry and by independent economists and analysts.

The output statistics provide data in greater levels of detail than ONS, on behalf of the UK, is obliged to provide to other international agencies, such as Eurostat. There is no statutory obligation for ONS to provide such detail, but there is significant demand from users for detailed breakdowns and regional measures of activity.

The data is provided in current prices (cash terms) and on a volume basis where estimated deflators are used to adjust the current prices to remove the effects of inflation. The volume measure is intended to provide a like-for-like trend in activity. This is necessary if construction output is to map consistently within the national accounts. It is also of great benefit to users who would otherwise individually need to assess inflationary effects. The data is also provided in a seasonally adjusted form to smooth the effects of underlying temporal or seasonal variations.

ONS also publishes breakdowns of workloads by size of company and it models output by region, as the survey does not ask respondents to allocate the values of work done regionally.

In essence the construction output series is the benchmark data used to assess the state of the industry. It also provides a common set of statistics for policymakers and lobbyists and a baseline for construction forecasters, for example the Construction Products Association[5] and Experian.[6]

Definitions of construction output and the selection of firms for inclusion

Deciding what construction is and establishing a method for selecting consistently representative samples from a coherent population is the foundation on which the value of a statistical series is built. It is a far from easy task. The construction industry is highly heterogeneous, both in its organisational structure and the products it provides. It is very mobile and in constant flux and its activity is volatile. Furthermore, as it provides the so-called built environment, its activities spill into every aspect of the economy, so its boundaries are often fuzzy. Boundaries and definitions must be set and rules and processes put in place to determine its population and, as necessary, select representative samples. In the ONS document *Questions and Answers Regarding the ONS Construction Output Survey* (ONS n.d.[d]), output is defined as

> the amount chargeable to customers for building and civil engineering work done in the relevant period excluding VAT. As well as work charged to customers, businesses are asked to include the value of work done on their own initiative on buildings such as dwellings or offices for eventual sale or lease, and of work done by their own operatives on the construction and maintenance of their own premises. The value of goods made by businesses themselves and used in the work is also included.
>
> In all returns, work done by sub-contractors is excluded to avoid double counting, since sub-contractors are also sampled. Output does not include payments made to architects or consultants from other firms – this would also cover engineers and surveyors. It would include wages paid to such people if they were directly employed by the business.

The construction sector is defined as all companies classified to construction within the Standard Classification 2007 apart from 41.1 'Property developers' (see ONS 2007). The reason given for excluding property developers is that most work is contracted out to businesses which fall within the definition, so construction activity should be picked up by contractors and sub-contractors.

In practical terms the firms are selected from the government's Inter-Departmental Business Register (IDBR) (ONS n.d. [e]), the sampling frame used for all ONS business surveys. Before the new series (the first data is for January 2010), firms were selected from the Builder's Address File (BAF). IDBR provided a wider coverage as it includes PAYE-only businesses and was also structured according to the 2007 Standard Industrial Classification (SIC 2007).

Since 1948, when the SIC was introduced, the classifications have been revised in 1958, 1968, 1980, 1992, 1997, 2003 and 2007. Revision is necessary as new products and new industries producing them emerge, and to reflect shifts of emphasis occurring within and between existing industries. This is useful. But with change come issues of consistency over time. This would be unavoidable with almost any system one might imagine, but the implications and potential foibles of selecting populations using SIC codes must be understood if misinterpretation of the data is to be avoided.

The main source of data is the Business Register and Employment Survey (BRES), used to populate the IDBR from which ONS draws its samples. BRES provides details of local business units, activities and employment. From this SICs are assigned. The data from BRES is also matched to data from HM Revenue & Customs (HMRC) on value added tax (VAT) and PAYE and with Companies

House[7] data, but BRES will always take priority when coding. Data on newly registered businesses comes from HMRC.

For some businesses the appropriate SIC code may not be clear-cut. They may undertake a range of activities, of which construction may be a minor part, vertically or horizontally integrated into the business. Firms, for instance, may select up to four SICs for inclusion on their annual return to Companies House.

Within IDBR the ONS follows a general rule to assign an SIC to each business. The SIC is assigned according to the activity of the business that provides the greatest gross value added (GVA). To ascertain the GVA, which is not directly captured by ONS, a proxy would generally be used, usually employment. Where an assigned SIC changes, the ONS would follow up with the company to check its validity.

In its favour, the SIC system does provide a recognised framework and some consistency with the surveying of other business sectors, and it is tied to international definitions. SIC is based on NACE,[8] which makes it comparable not just across Europe but worldwide. That it is updated to reflect changes within the economy is also useful. But it is important to be aware of the problems that stem from using SICs as the test for inclusion and exclusion, the implications of which are examined later.

It is worth emphasising that the construction output series covers just Great Britain; it does not include Northern Ireland, the Channel Islands or the Isle of Man, so on that basis it is covering a narrower geography than the UK GVA measure.

The implications of the definitions and selection

The ONS applies a statistical disclosure-control methodology to avoid identifying individuals or individual organisations. The constraint created by confidentiality means that some of the following discussion on the formation of the population of construction firms is speculative. In the absence of verification, the aim here is to suggest where issues may lie to illustrate the challenges in creating a construction output series and highlight potential foibles.

Inevitably the ONS has to draw boundaries between sectors if it is to generate consistent data. However, most big firms carry out a range of activities, mainly in support of a main activity. This support work may include activities such as administration, accountancy and finance, human resources, information technology (IT) and facilities management. Equally, firms may choose, as many have increasingly over recent decades, to outsource non-core functions.

It is impractical to unpick each single activity within each and every company in an attempt to assess accurately how much accountancy, IT support, human resource management or construction is being undertaken. It is also exceptionally hard to correct for changes in the level of non-core activities undertaken directly or outsourced. Therefore, if all firms rapidly outsourced one function – say the accountancy function – to dedicated accountancy firms, we would see a spike in activity in that sector without necessarily any rise or fall in the overall level of accountancy work taking place.

From the perspective of most people who work within construction, the ONS definition of construction implied by the inclusions and exclusions is likely to be regarded as narrow. A significant amount of the value added in creating the built environment is missing, notably the planning, design and many other necessary professional services activities, although these are included under other sections of the SIC. Certainly, it might seem strange even to those outside construction that the input from architects, quantity surveyors, planners and civil engineers not working for contracting firms is excluded from measured construction. And, it must be remembered, construction work undertaken in-house by non-construction firms is also not included.

While the ONS definition may be narrower than that which most people might regard as the construction industry, it is conceptually different and wider than the GVA contribution from construction used in estimating the nation's GDP, as the materials used in the construction are included in the amount charged to customers.

It might be reasonably assumed that the GVA measure that feeds into the UK GDP preliminary estimate is a measure of the amount of value added in the site assembly process; however, while SIC coding may be sophisticated, in reality, and certainly at the margins, it can prove relatively crude as a consistent means of sorting construction and non-construction activity. Both the construction output and the GVA estimate will miss construction work carried out by non-construction firms and include work that might otherwise be considered architectural or consultancy work or manufacturing work that is done in-house, but off-site, by contracting firms.

Large organisations that may have sizeable in-house maintenance and construction operations, such as housing associations or utilities, will be excluded unless those operations are set up as companies that are coded as construction within the IDBR. There will be many businesses undertaking significant amounts of construction work while it is not their main activity. In addition, there will be work undertaken by individuals as self-build which will not be captured within measured construction. However, particularly in house building, construction described as self-build may in reality call on the services of firms that are classified as construction firms. This creates a further grey area when interpreting the measurement of the industry.

Major changes to the National Accounts in 2014 included the treatment of own-account construction, which includes construction by non-construction companies, self-build new housing and improvements (ONS 2014b). The impact of this change was large, adding GBP4 billion to GDP in current prices for 2009. Within the calculations an estimate of GBP2.5 billion was given to the own-account construction by private non-financial corporations.

Related to the issue of what is included within the definition of construction is an important change in the population of construction organisations covered following the switch from BAF to the IDBR database, when the new ONS series was introduced. Local authority direct labour organisations (DLO), which had been included in BAF, were excluded. The case put by ONS was that output from DLOs was falling and, at the time, was estimated to contribute less than 2% of overall construction output. Meanwhile, a perceived improvement was that the wider coverage of IDBR meant that more firms (notably micro businesses) were captured. Previously the output data had included an estimate for unrecorded output to account for firms not picked up by BAF. This unrecorded output estimate is not part of the new series.

It is worth noting the difficulties in measuring unrecorded output and why seeking to measure it may create unforeseen issues and misleading data. The pre-2010 measure of output based its estimates on the difference between the employment captured by the Labour Force Survey (LFS), including self-employed, and the employment measured on its database of construction firms. This tended to be around 600,000 persons. Wage information and average output per person for smaller firms was then used to estimate the output from the missing workers. However, in the period around 2004 there was a large influx of migrant workers. For practical reasons, the LFS has a tendency to under-record migrant labour as many migrants initially live in houses in multiple occupation, which tend to provide poor responses. The likelihood is that as migrant labour in construction increased, the estimate of unrecorded output was lower than it otherwise would be, as migrant labour would be expected to be more heavily concentrated among the self-employed. This has been suggested as one possible explanation for an unexpected trough in the official estimates for repair and maintenance work from 2003 to 2008 (Green 2010).

Businesses will inevitably migrate between the sectors cast within the IDBR, as the balance of their work or employment changes and as codes are revised. This leads to shifts in the population of firms, and theoretically the potential for a very large firm with a mix of work close to the borderline between two activities to be switched from one classification to another as a result of a relatively small change in the mix of activities. The effect could be quite dramatic.

Furthermore, as mentioned above, the approach used to allocate SIC codes uses employment as a proxy for GVA to segment the activity within firms. The firm is given the SIC code associated with its largest activity. This has limitations, especially where the labour intensity of the activities varies greatly within a firm. There is an implicit assumption in using employment as a proxy for GVA that, on average, the workers in each activity are equally productive. In reality, low-productive labour-intensive activities within a given business will have a disproportionate effect on determining the SIC code allocated. For firms with activities spanning different SIC codes, changes in employment in the more labour-intensive activities will have a disproportionate effect in determining which SIC code is allocated to the overall business.

Reclassification of businesses with the IDBR can have a profound impact on output figures. In the August 2015 data (released in October), there is what amounts to a structural break created by a services company being reclassified as construction. The effect was to increase output by around 1%, an estimated GBP1-2 billion a year (Green 2015).

There is also significant room for ambiguity. This is evident within plant hire. If we look at how firms are SIC classified in Companies House data, some plant-hire companies are SIC 77320 ('Renting and leasing of construction and civil engineering machinery and equipment'). This would place them as a service sector company. Others are SIC 43999 ('Other specialised construction activities not elsewhere classified'), which would include them within the population of construction firms. Under SIC the difference is whether the plant-hire firm supplies an operator or not. Renting with an operator is classified as construction; without is classified as services. Many firms may provide both in varying amounts and some, in addition, may undertake significant construction work such as groundworks.

Whatever the value of their activities, the allocation will depend on the employees assigned to each activity undertaken. This leaves a question over how a plant-hire firm might be classified if it undertakes significant construction work using sub-contracted operators, while it uses direct labour for its non-operator rentals. In a similar vein, ambiguity might also arise with haulage firms that specialise in construction and undertake what might be regarded as groundworks. The central point here is that in many cases differences in activities between construction classified and non-construction classified firms may be very subtle and not picked up neatly through the SIC assignment process.

The challenge is not simply about seeking to measure actual construction activity accurately at a point in time. For firms on the cusp between activities, relatively small shifts in activities can tip them into a different sector. How this might influence trends in the published statistics is hard to estimate, but inevitably, at some level, it does. Historically, contractors owned significant amounts of plant and equipment. Today they tend to hire. So we are not measuring the same construction industry today as we did in years past.

There is a more important transformation of the corporate structure and the nature of work undertaken by construction-related businesses that has taken place over the past two decades or so. Again, how this impacts on measured construction is hard to quantify. In the 1990s, possibly prompted by a deep recession, there was a major shift in thinking among major stock-market listed construction firms. Part of the attraction would have been the higher price-earnings ratios within the services sector of the stock exchange compared with construction. However, the shift reflected a profound change in business strategy and a response to the rising interest in facilities management and the Private Finance Initiative (PFI). A number of major contractors, such as Carillion, Interserve (formerly Tilbury Douglas) and Amey, redefined themselves as support services businesses, covering an array of activities. Even contractors, such as Kier, who did not seek to redefine themselves as service providers established divisions that included diverse activities, from waste collection and recycling to social housing maintenance.

This trend was, however, not limited to publicly listed major contractors. WS Atkins, a construction-based consultancy with a long pedigree, diversified into contracting-based support services such as highways maintenance contracting. Other firms emerged rapidly, for example Enterprise, Accord (demerged from John Doyle Group), Connaught and Mitie, that sought to win market share in the expanding market for outsourcing non-core activities, in part accelerated by compulsory competitive tendering legislation under the Local Government Acts of 1988 and 1992. The effect of this structural reorganisation of businesses engaged in designing, creating and maintaining the built environment has been to further blur the distinctions between roles and activities.

For most of the construction firms that migrated into these activities, much of the work remained, to all intents and purposes, construction, but the balance of the contractual relationships and the scope of work did change as firms expanded their service sector work. How this affected measurement of construction activity at the time, how it may have altered the pattern of firms within BAF, and what effect it might have on those that are now classified as construction within IDBR are hard questions to answer without very close scrutiny of the micro data. It will, however, have pushed more firms closer to a boundary between the services sector and construction. In doing so, this will have increased the likelihood of large firms switching between the two. As mentioned earlier, the reclassification of a single services firm into construction had a profound impact on the construction output series.

The exclusion of professional services firms, such as civil and structural engineering, quantity surveying and architectural practices, presents its own set of issues when seeking to interpret the consistency of the construction output time series. In recent years, especially with the increase in design-and-build contracts, contractors and sub-contractors expect to undertake increasing amounts of the design work that previously would have been undertaken by consultants employed directly by the client. Today it is common within contracts that the services of the professions are novated to the main contractor. This work would not be collected as construction output, but how this shift in practice has impacted on the level of design work undertaken directly or contracted out by contractors and how this may have impacted on the measure of construction output is unclear.

The exclusion of firms classified as SIC 41.1 ('Property developers'), meanwhile, presents a different problem. Excluding this classification is not trivial. According to the Annual Business Survey, the estimate of GVA by SIC 41.1 in 2015 was 12% of all GVA for construction (ONS 2016). These firms only became part of construction when SIC 2007 was introduced; previously they had been included in 'Real estate'.

House-building firms fall awkwardly between developers and builders. We see this in the SICs on the Companies House register where, for example, Taylor Wimpey Plc is coded 70100, while Taylor Wimpey UK is coded 41100 ('Development of building projects'). That code is also used by Persimmon and Bellway Plc. However, Bellway Homes Limited is coded 41200 ('Construction of domestic buildings'), as is Redrow and Barratt Developments Plc. Bovis Homes Group and its main subsidiary companies appear on the other hand to be coded more consistently as 68100 ('Buying and selling of own real estate'). These firms would be regarded as being very similar in nature and operation. Again, without looking in detail at the IDBR, it is unclear how each of these similar businesses is classified for the purposes of construction output when the ONS proxy of employment is used. It is worth noting that ONS practice is to not disclose the SICs of companies, and there may be inconsistencies between what is recorded at Companies House and what is recorded at ONS.

A common feature of all major house builders is that they sub-contract out the majority of the actual construction work, but most do significant construction-related work and directly engage teams employed in site management, skilled trades and other production functions that might reasonably be expected to fall within the scope of construction output. There is then likely to be a rather blurred and wobbly line drawn between property developers and construction firms, a line that is further confused because some contractors also undertake speculative house building.

Attempts to include or exclude property development raises other questions. For instance, with regard to house-building firms, it is unclear how accurately they (or indeed contractors undertaking speculative house building) might separate the value added through construction processes and trading in land. This is particularly pertinent to the measurement of the margin made on speculative sales, which for house building is generally far greater than for contracting. It also tends to be far more volatile. If developer margins are used, there may be a significant overstatement of the work done in private housing.

Looking beyond more conventional construction-related firms, such as house builders and contractors, the increasingly complex nature of buildings and structures is drawing in skills and businesses that may not have been used in the past or may not have been used as widely. This, too, makes measurement of construction output on a consistent basis more difficult. For instance, buildings and infrastructure are increasingly adopting so-called smart technologies to increase efficiency. Whether these are installed by a business assigned an appropriate SIC code or not will determine whether its work is included with, or excluded from, construction output.

An example of ambiguity created by the changing nature of construction-related work might be the installation of a wind generator. Does the whole of the wind generator and its installation fall within construction, or just the installation or neither? This may vary according to the contract and the contractor that undertakes the work. If the generator is supplied by a construction firm, it may be that all the output is assigned to construction output. If a manufacturer undertakes the installation and related construction directly, none might be included. If the client buys the generator and installation separately, it may be the installation work that is classed as part of construction output.

The point is less whether or not and to what extent this work should be included; it is the uncertainty that is bothersome. It is plausible that for similar work different measures of construction output might be logged. These complexities are being looked into, but at the time of writing there is no detailed answer to how this issue might influence the level of measured construction output and what solution might be adopted if the effects are significant.

Further uncertainty results from the establishment of joint-venture businesses, which appear to be increasingly popular in construction. First, these may often be temporary and short-lived entities, so they may be missed. Second, they may well not be SIC coded to construction. In many cases they may simply be administrative vehicles directly undertaking no construction work. It is the uncertainty over what construction work is or what might not be finding its way into the output data that presents the problem, especially if the popularity of joint ventures fluctuates greatly over time.

Considering all of the above, it is evident that significant shifts in measured construction activity can potentially occur without underlying change in real activity. Shifts in the mix of work and corporate activity may have a significant impact on the amount of construction activity measured. Perhaps more importantly, what may appear to be trends in activity may in part be more rooted in trends in corporate strategies or contractual relationships.

Looking beyond the output data to the measure of GVA, further issues arise. Currently the Great Britain (GB) construction output measure provides the basis for estimating UK construction GVA. But the two measures are different geographically. Furthermore, the output measure includes the value added embodied in the materials used while GVA does not. And the value of construction delivered by property developers is not covered within construction output, but it is within GVA.

Currently a single factor is used to make the adjustments, both scaling down for the extraction of materials and scaling up to include property development construction activity and construction work in Northern Ireland. In effect GB construction output is deemed to be 33% larger than UK construction GVA. While this provides a simple method of scaling, it inevitably introduces errors within the time series.

This rather simple factoring implies assumptions that can be questioned. It assumes that, on balance, the ratio of UK construction GVA to GB construction output will remain constant. However, the materials and labour content will shift greatly as the mix of work shifts. More obviously, we know that construction output in Northern Ireland has declined markedly in recent years relative to Great Britain. In 2007 output in Northern Ireland was about 2.7% that of GB; in 2014 it was about 1.6%. Assuming all other things are equal, the factor would overstate UK GVA in 2014 relative to 2007 by about 1%.

Joint ventures will be included if they fall within normal selection parameters. Subsidiary companies will be included in the sample depending on how the business is structured for reporting purposes. Holding companies are not included and it is unlikely that dormant and non-trading businesses will be included, although there may be a lag in some of these disappearing from the register.

Collection of data

At the time of writing, the data is collected through a monthly paper-based survey sent by post to about 8,000 firms sampled from the IDBR. Returns for businesses that do not respond are imputed and the data is weighted to provide estimates for the full population.

The sample is taken from about 230,000 firms deemed, within IDBR, to be construction businesses. The size of these firms varies greatly, so the population is stratified to increase the efficiency of the surveying. All firms are included that employ more than 100 or employ between 10 and 99 and have annual turnover GBP60 million or greater. Random samples are taken for firms employing 0 to 4 employees, 5 to 9 employees and 10 to 99 employees with a turnover less than GBP60 million. ONS analysis suggests the sampling fraction in each sampled stratum approximates to optimal as determined using a Neyman allocation.

It is a fairly simple survey which initially asks for the reporting period, if it is not the period stated in the questionnaire. It then asks for the value of "actual work carried out on housing, infrastructure and non-housing". There is an additional question on employment and a panel for additional comments and contact details, allowing for firms to report that they have done no work.

The survey collects data on values to the nearest £1,000 for work types undertaken and chargeable to customers in the reference period in England, Scotland and Wales. Firms are asked also to include work done for themselves either for later sale or by them on their own premises. It is a GB survey, not a UK survey; the survey does not cover work in Northern Ireland, Channel Islands, the Isle of Man or overseas. The inclusions and exclusions presented in the questionnaire help to understand what is being measured and how measured construction output might change with changes in how the industry operates.

The survey asks respondents for values that include the following:

* Work on buildings which you hope to sell later for profit (speculative work);
* Work done by your business on its own business premises;
* Fixtures, equipment and tools your business made and used in construction; and
* Materials your business used, overheads and profits and labour costs on your payroll.

Excluded are:

* Payments made to sub-contractors;
* Payments to architects and other consultants;
* Fixtures, equipment and tools your business made for sale;
* Materials your business sold;
* Value of land; and
* VAT.

As mentioned above, this tells us much about what construction output is and how structural changes to the industry can, without any more or less work being done, influence the level of work recorded.

The survey itself collects 11 values for work sectors, a total value and further information on employment. The 11 sectors are:

* Private new housing;
* Public new housing;

- Private housing repair, maintenance and improvement;
- Public housing repair, maintenance and improvement;
- New infrastructure work;
- Infrastructure repair and maintenance;
- Public non-housing, non-infrastructure new work;
- Public non-housing, non-infrastructure repair and maintenance;
- Private commercial new work (including improvements and extensions);
- Private industrial new work (including improvements and extensions); and
- Private non-housing, non-infrastructure repair and maintenance.

In questionnaire design there is always a trade-off between striving for more detail and ensuring a simplicity that reduces ambiguity and increases the chances of accurate completion. In this light the balance achieved within the construction output monthly survey seems appropriate.

It should be noted that finer detail provided in the construction output release and any regional breakdowns of work done are modelled, as actual details are not collected.

What is not easy to gauge, without more detailed surveying, is how accurately the forms are being filled in, given that there are potentially awkward calculations expected of at least some firms if they are to ensure that they comply with the specified inclusions and exclusions. Anecdotal evidence from larger contracting firms that respond to the survey suggests that the person delegated to undertake the task is often junior. This in itself is not necessarily an issue, although where discretion is needed they may not have all the necessary information.

One example might be the exclusion of the value of land. This requires a cautious approach. For normal contracting work there is not an obvious issue as the land is not part of any transaction. For speculative house builders or the speculative house-building arms of contractors that are surveyed, accounting accurately for the land may prove awkward. Questions raised might be: What value of land should be excluded from the price paid by the buyer? What proportion of the margin is attributable to property development and what to construction? Similar questions might apply in how construction projects undertaken as Public Private Partnerships (PPP) or under the PFI are accounted for.

While the separation into public and private is deemed useful by those seeking to interpret trends and understand the drivers of construction activity, the likelihood is that the data provides limited clarity. There are many factors that might cause confusion. First, the person filling out the survey may also be unclear, especially if they are a lower-tier contractor. They may be one or more steps removed from the ultimate client. Their immediate client is likely to be a private contractor, but they may not be certain whether the work is for a public or private customer. It seems unlikely that all responses will be correct, and without more detailed evidence it is hard to say which way a bias might lie.

Furthermore the distinction between public and private housing has become increasingly unclear, with private contractors delivering housing to housing associations that should be classed as public housing, and with housing associations building homes for sale, which should be classed as private housing. And, even if it were, work that may have been started in the private sector could be real-located to the public sector.

It may seem obvious how "work done" is carried out over the reference period, but it is not as simple as it seems and data may not be readily available to those filling in the questionnaire. For convenience some firms state the value of orders received, some state invoices raised, some state turnover or invoices paid, while others provide an estimate of actual chargeable work done in the reference period. This means that measured construction output in reality is a mix of slightly differ-ing measures, with the result that the received data has an unknown and variable lag if it is actual work done that we intend to gauge.

In December 2011 the ONS published a small analysis of this problem after suggestions that this may disguise the timing of the impact of major events, such as severe weather, within the time series (ONS 2011). The majority of firms ques-tioned were found to state invoices raised as their measure, but the majority of larger firms estimated work done. The results from the relatively small sample of 187 firms did confirm that there were variable lags and suggested that the effect was likely to be greatest among work recorded by smaller firms.

There was some criticism of the conclusions of the analysis (Green 2011). The impact should be considered, especially where a major event prompts a slowdown in activity late in one month. For instance, if very heavy snow in late January dra-matically lowered activity on sites, a significant effect from this would be seen in the February figures, producing a much lower measured output relative to actual work undertaken.

The response rate for surveys within the month is around 70% and, with larger firms tending to return surveys more promptly, the coverage of output tends to be more than 75% and generally nearer 80%. There will naturally be late returns providing new data that is added to the existing data, displacing imputed data. This prompts revisions, applied within the previous 12 months.

Statistical treatments of the survey data

Given that the survey takes stratified samples and not a complete census, signifi-cant work is done to process the data taken from returns to provide a reflection of actual work done. It is worth noting that no treatment is used to adjust for any lag effects inherent in the returned data, as mentioned above, that is caused by firms using different measures of work done over the reference period.

Returns are scanned for anomalies or unusual returns. This may prompt ONS to contact a firm to check the details, potentially resulting in corrected data being resent, or some may be deemed correct and then treated as outliers.

One of the great powers of statistics is the ability to treat sensibly anomalies within the data from samples of a population so that they do not overly influ-ence the general pattern. Ironically, for many analysts and observers it is often the quirks that are most interesting, providing in some cases early indications of turning points in a trend or other emerging trends. However, leaving outliers within a sample that is to be grossed up risks overestimating the overall total. The approach taken with the construction output statistics is the one-sided Winsorisation method (Lewis 2007). The method involves identifying thresholds. Any values lying above the thresholds are reduced towards the threshold. The parameters used are periodically reviewed and updated, as they were in late 2015 (Davies and John 2015).

Concern over a possible understatement of activity in the early months of each year led to an investigation which found higher numbers of outliers occurring in the first four months of each year. Changing the parameters resulted in the average of outliers per month falling in these four months from 70 to 17. The effect was to raise recorded output in January, February, March and April by a monthly average of about GBP166 million. The subsector which saw the greatest fall in outliers was private housing. This reclassification of outliers inevitably has a knock-on effect on the seasonal adjustment.

For businesses that fail to respond, even after chasing, data is imputed, with imputation values estimated from the pattern of responses from similar business. A link factor is derived which is applied to the previous returns for the non-responding firm. If data is returned subsequently, the imputed data is replaced. For a firm that has never responded, an original value is calculated from a ratio of the registered turnover of the firm, the ratio being derived from the returns and registered turnover of other firms.

The sample data are weighted to provide an estimate for the full population. Here ONS uses two weights: a design weight that is the inverse of the inclusion probability of the businesses in the sample, and a calibration weight to adjust for the imbalance of the selected sample compared with the population from which it was selected.

The published monthly releases do provide an estimate of the current price value for work done in a month. But users appreciate adjustments made for seasonal effects to provide a clearer view of the underlying trends, and both the current and the volume measure provided are seasonally adjusted using X-13ARIMA-SEATS developed by the US Census Bureau (US Census Bureau n.d.). It is important to note, however, that as it is a new series and also because construction is a highly volatile industry and the series began during a major recession, the seasonal adjustment will generate revisions of a greater scale and more frequently than might otherwise be the case. The relevant parameters for construction output are reviewed annually and by time-series analysis experts.

The treatment of the data that has created most concern is deflation, as mentioned above. The aim of deflation is to provide a series that approximates to the volume of work done after inflation effects have been removed. This is extremely complex and will not be covered in detail here. Issues such as capturing improvements in quality or positive and negative externalities generated by changes in regulation or practice (for example health and safety or environmental improvements) present major problems but are generally ignored.

Until recently, price and cost indices used to adjust for inflation were provided by BEIS. The original indices relied heavily on data taken from bills of quantities. With changing procurement patterns these have become used much less frequently and the samples from which they were taken became too small (Yu and Ive 2015; UK Statistics Authority 2011b).

In 2013 BIS (now BEIS) awarded AECOM, a construction consultancy firm, a contract to supply cost and price indices. Part of the contract was to investigate and, if possible, develop a new approach. A new model-based approach was developed for use in the December 2014 publication (Q3) (Gov.uk 2014), but BIS took the decision not to publish this release because of concerns about the outputs from the new approach. This coincided with the UK Statistics Authority suspending the National Statistics designation. Concerns persisted. The responsibility for construction price and cost indices was taken in-house by ONS in April 2015 (Gov.uk 2015). The ONS decided to develop an interim solution for construction output price indices. These were first used in the release published in June covering output up to April 2015 (ONS 2015b).

Continuing concerns led ONS to constitute a construction prices steering group of key users to assist in the development of a solution for the price indices. This work has been widened to a more detailed review covering all aspects of the construction output statistics.

Over the past five years or so there has been significant volatility in the published data. This volatility is likely to continue as ONS seeks to put the construction output series on a firmer footing. Certainly, the issues around the deflators and other issues that have emerged provide a clear illustration of the complexity inherent in attempting to map construction activity over time on a like-for-like basis.

Data outputs and modelled data

As shown above, the construction output survey collects a limited amount of hard detail. It provides data on very broad work categories and within limited employment bands. It provides scant evidence for where work may be taking place geographically, as construction occurs at multiple unrecorded locations away from the contractors' established bases of operations from where the data is collected. To provide users with indicators of workload in finer detail, ONS models data to provide regional breakdowns and finer splits of the subsectors. This is provided on a quarterly basis.

To estimate the splits of new work, the ONS uses orders data. This is now provided under contract by Barbour ABI (ONS 2014c). This provides a value of orders at postcode level and by type. However, the flow of each order into a stream of construction work (output) is not straightforward, as the actual workload will be spread over time, which will vary contract to contract. To account for this, the model includes assumptions based on the previous PROBE (quarterly inquiry projects in progress) survey, which was carried out by BEIS. In effect, orders values are projected forward and the totals for each quarter in each region and by each category are estimated. A similar process provides estimates for the output in subsector categories.

The orders series does not, however, cover repair and maintenance. With less evidence available, it has been assumed the repair and maintenance work is conducted in the region in which the firm surveyed is based. This assumption may hold reasonably for smaller firms and a certain amount of cross-border activity will be cancelled out, but larger firms will undertake work within a far greater radius of the administrative office

There are clearly significant shortcomings with this modelled approach. New orders may be cancelled and thus the work will not be done. The value of work carried out under any given order will vary at differing rates; it may shrink or expand above the original value. The pace of work will vary differently to that projected. For repair and maintenance, the shortcomings are greater. Furthermore, if a large contractor moves offices to another region, the workload would follow creating a jump in the time series. These failings are recognised, but users are keen to have some, albeit shaky, data to this degree of fineness.

Future prospects and challenges

The collection, analysis and dissemination of statistics does not happen in a vacuum free from financial, organisational or political pressures. The mere decision of the State to collect data (from which the word statistics is derived) is ultimately political. The choice of what statistics to collect and in what detail will rest on many factors other than the demands of users or their economic value. The public resources made available for the collection and analysis of statistics will be a political choice and there will be obligations, statutory or otherwise, to national or international bodies, such as the European Union, UN or OECD, which will have a bearing on decisions made. The perceived costs and benefits will be weighed between the potential administrative burden on business and the benefits they receive from the data. The longevity or embeddedness of an existing series will also have a bearing on what is collected and how. There will be path dependency, and more relevant or useful complementary or competitive data series or methods may struggle to become accepted, despite being more fashionable and having greater support from users and potential users.

It will not be simply the usefulness or cost effectiveness of data collection that will shape what is measured, how it is measured and what data is collected, analysed and presented; there will be a range of political, economic, social and technical considerations, along with other often more specific factors. And the weight given to each consideration will inevitably vary over time.

Three current influences that appear to be in the ascendency in respect of their potential impact on the collection and dissemination of statistics, including the construction output series, are worth closer consideration. The first, the most obvious and intriguing, is the rise of digital data generated in vast and increasing quantities since the development of the internet. Second, there is the political desire to reduce public sector costs. This means funding for statistics is more likely to be squeezed than loosened. Third, there is increasing desire to reduce red tape and the burden on industry, in part connected to a more general desire to increase productivity.

How much these factors will eventually influence how statistics are collected, analysed and disseminated is unknowable. The impact is likely to depend as much on political persuasions, whims, opportunism or fashion as it does on any evidence or evidence-based evaluations.

Looking at the combination of these three factors, it is quickly apparent why in recent years there has been a growing interest in using administrative data to support or supplant more traditional approaches such as direct surveys. Administrative data is collected by or about people and businesses for a host of reasons, both statutory and commercial, by a variety of organisations. Furthermore, the increased digitisation of administrative records that has run in parallel with the growth of the internet has made it cheaper and quicker to collect and manipulate the data.

The UK government has great enthusiasm for using administrative data. A Cabinet Office letter to the UK Statistics Authority states its ambition and heralds the end of the paper-based decennial census of population. It says:

> our support for the dual running of an online (decennial) census with increased use of administrative data is only relevant to 2021 and not for future censuses. Our ambition is that censuses after 2021 will be conducted using other sources of data and providing more timely statistical information.
>
> (ONS 2015c)

Looking to businesses, administrative data appears to have greater attractions than population surveys. This approach has the potential both to reduce the resources used and costs associated with the generation of statistics, and to reduce the burden on businesses. This would be particularly true of small firms, where relative to their size, form filling would greatly increase the proportion of productive time lost to completing surveys. Administrative data will not be without errors, but it removes many of the errors that occur when respondents are filling in occasional surveys, and there are generally in-built correction processes, for example, correcting for overpayment or underpayment of taxation.

Considering construction output data, how might access to administrative data help? There are numerous repositories of data related to construction, such as employee, employer and sales tax records. Firms themselves keep electronic accounting records covering money in and money out which track the functioning of their businesses. For instance, creating software applications that automatically generate survey data through extracting data from a firm's management information systems could lead to lower burdens on business, reduced surveying costs, greater accuracy and larger samples of smaller firms. This could, in turn, lead to the capture of a richer and wider seam of data and provide enhanced understanding.

In the case of construction output, for instance, this might enable a better insight into actual prices paid and costs incurred, in addition to a more robust understanding of the actual work undertaken. With potentially lower cost and technical barriers, data from other sources could be exploited; for example, details on materials prices or activity could be obtained directly from merchants or other suppliers.

The challenge is to find ways to capture this data that are secure, respect confidentiality and help describe the functioning of the firms and the industry overall. Such developments may be iterative in that they capture records electronically and, to a greater extent, automatically mimic the current survey approach or enhance the richness of the surveying process. But the potential is there for radical development, taking advantage of new approaches to reconfigure the way business activity is measured.

It is not simply business administrative data that could add to the understanding of construction output. The advent of Building Information Modelling/Management (BIM) also has potential to provide a vehicle for greater understanding of the industry which might inform the measurement of the industry, if only by enriching the understanding of relationships between parties.

It is inevitable that the use of administrative data will shape how construction is measured. The question of how that will occur is yet to be answered.

But the vast quantities of data generated as a result of the internet explosion, and the ever increasing analytical power of computers, have brought other potential, and indeed conceptually very different, ways to observe the world. Described as "big data", huge amounts of often disparate data are being pulled together and analysed to identify patterns which can help us better understand the world in which we live and the many ever-changing aspects of that world.

The advent of big data is revolutionary in that observations or actions previously ignored and not recorded, or not recorded in an accessible way, are now gathered and interpreted. It relies on connectedness. While the term is widely used and prone to ambiguity, Laney (2001) has suggested that the three dimensions of big data are generally regarded as being volume, velocity and variety. While Laney was considering these three dimensions in relation to data management, they provide a framework for appreciating the growth of big data. Data is being collected on more actions and transactions, and more detail about each is being collected digitally. This is increasing the volume. The velocity is increasing, with an ever-shorter time between an event occurring and the opportunity to measure it. Also the variety of data which can be captured, compared and analysed is increasing as previously incompatible and unconnected data sources are connected and provided in, or converted to, common data formats.

Its application at present in measuring construction output is limited. But it can be used to tell us about relationships that might help us interpret changes and test existing data. It can also be used as a predictive tool. It opens up methods of analysis that might hitherto have been dismissed. Having vast quantities of data can reveal sometimes obscure patterns. So, for instance, it can be used to measure change in interest or activity through textual analysis that measures the occurrence of 'buzzwords'. At the most basic, one might note the results returned by Google for a given search term for a given period at various points. This would tell us something about the world. The value of that 'something' would lie in the relatedness of that search term to what it is we are seeking to understand, and whether correlations we find are spurious or have a causal link.

Such methods have been used with success in predicting election results and they are being tested for forecasting construction output prices and the like, combining big data analytics and data mining with behavioural economics and econometric modelling. It is as yet uncertain how well such an approach might develop beyond measuring change in the level of activity and be used in measuring the actual level of activity to fit the needs of users of construction output data.

In understanding the potential for big data techniques we also have to appreciate the rapid expansion of the Internet of Things. Smart devices, ever more pervasive, measuring numerous outputs offer huge potential in helping us describe the world statistically, directly or by proxy, and not least in regard to construction.

However, for all the technical and technological challenges that face the measurement of construction output, there will always remain two central questions. First, why are we measuring construction output? Second, are we measuring the right thing?

Path dependency will always be at play here. There is much invested in how we currently measure construction output and change will come at a cost. However, there are well-recognised fundamental concerns with the measurement.

Professor Charles Bean has undertaken a major review of the UK statistics. His report highlighted the errors and deficiencies in the construction statistics in recent years, but the report itself raises more fundamental questions. He said in the press release launching his report:

> We need to be candid about the limitations of UK economic statistics. The UK was one of the original pioneers of national accounting. We need to take economic statistics back to the future or we risk missing out an important part of the modern economy from official figures.
>
> (Gov.UK 2016)

This statement comes at a time when there is growing concern over the value of the GDP measure. In 2016, *The Economist* stated "GDP is a bad gauge of material well-being. Time for a fresh approach" (*The Economist* 2016). Fundamentally GDP was designed in an age of manufacturing. Today, in developed economies, services dominate and this shift in many ways echoes Maslow's hierarchy of needs. In line with growth in services there has been a shift in emphasis from consuming 'things' to consuming 'experiences'. Valuing the latter over time poses far greater challenges.

For the measurement of construction output, the concerns over GDP have a number of implications. First, a core reason for measuring construction output is that it provides data for the calculation of GDP. But, second and potentially more importantly, the nature of buildings and their value is undergoing rapid change and the shift in emphasis from things to experiences can easily be underestimated. The shift of public procurement from buying works to buying services (which could be regarded as experiences or outcomes) through PFI or PPP contracts is in many ways an illustration of this change.

As the performance of buildings improves and their value to users improves, as they embrace more smart technology and provide better experiences for users, the measurement of construction output as a site-based activity becomes more distanced from the value added in creating or maintaining the built environment.

This leads to a third related concern. Statistics underpin evidence-based policy. The separation between the statistics of construction output – what happens on-site – and the rest of the value added in creating and maintaining the built environment can lead to misunderstanding among policymakers. This is very evident in the understanding of productivity – currently a hot topic internationally. It is quite possible that we could create the built environment more productively, with greater emphasis on off-site manufacture and design, yet, as things are currently structured, the productivity of construction could fall. This was a point made in a recent report from the Chartered Institute of Building (Green 2016).

Where the industry has for many years sought to engage more collaboratively across boundaries to improve its performance, the results of this are to a great extent absent or disguised in the statistics. The lack of integration in the statistics presents a problem in analysing the overall performance of what ultimately matters – that is, how well we create and maintain the built environment.

Summary and conclusions

As we have seen, construction output measurement in the UK is in flux. In December 2014, the National Statistics designation *construction output statistics* was suspended and subsequently the Office for National Statistics has been investing in improving the series. Tough questions are being asked about its fitness for purpose, but the value of having detailed data measuring the activity of construction is not in question. This all comes at a time when there is great interest and potentially great change in how statistics generally are collected and analysed. It also comes at a time when information technology offers the hope of improvement and the possibility of new ways to measure construction output.

However, one of the most testing issues, one that will not be solved simply by technological advances or new statistical techniques, is the definition of the industry and what constitutes construction. That 'construction' is defined as the activity on sites increasingly presents problems. The challenge is likely to grow as the industry is encouraged to become ever more integrated across the divides between design, manufacture and assembly. The latter is currently regarded as the construction activity for statistical purposes; the other activities fall into services and manufacturing.

Not only does the lack of more complete and integrated statistics hamper the understanding of the wider construction industry, but it also creates major hurdles to making sub-sectoral or international comparisons. The balance of the inputs into design, manufacture and assembly, used to create and maintain the built environment, will vary country to country, circumstance to circumstance and over time, so like-for-like comparisons are at best difficult as well as potentially misleading.

One suggestion gathering momentum is the notion of creating satellite accounts. This is a term developed by the United Nations to present data related to economic sectors that are not defined as industries in national accounts. Such data series are created internationally for tourism under the auspices of the United Nations. Meanwhile, the UK collates data for the creative industries and the creative economy. This data, interestingly, includes architecture along with museums and the performing arts (WTO 2008).

The idea that ONS should give greater consideration to satellite accounts is not new. The Atkinson Review carried out for the National Statistician and published in 2005 recommended that "ONS should explore ways of analysing and publishing information about public service outputs in parallel to the National Accounts, such as satellite accounts. In particular, it would be useful to have a satellite account on human capital resource formation" (Atkinson 2005: 100).

The aim in relation to construction would be to create a set of accounts that represent the delivery and maintenance of the built environment. These would capture the output, the full value added, the full level of employment and other key measures of what is generally regarded as the wider construction sector.

The creation of satellite accounts would not be simple, nor would it remove many of the difficulties of capturing the industry statistically. It will remain volatile, heterogeneous and mobile. But the introduction of satellite accounts would better capture the value and changes of a sector in flux and ultimately provide a better array of statistics with which to gauge its performance and progress.

Pearce (2003) made this point on construction data:

> It is clear, however, that different sources produce different results and can generate, at best, uncertainty and, at worst, confusion. The main reason for this is that construction is not a tidy industrial sector; it does not fit comfortably into the three basic industrial categories of primary/extraction, secondary/manufacturing and tertiary/services. Construction is an assembly industry. It takes goods and services from other industries to produce its product, the built environment.
>
> It is not possible to change the national statistical system for the convenience of the construction industry. It is possible, however, for the industry to influence how data from official, industry, and other sources is collected, analysed and presented.

This still holds today for the measurement of construction output and, if anything, is of greater importance.

Notes

1 In addition to the official construction output data series there are other sources that gauge changes in industry activity, directly or indirectly. Trade and professional bodies,

such as The Construction Products Association, Federation of Master Builders, Build UK, Builders Merchants Federation, Civil Engineering Contractors Association, RIBA, RICS, provide state of trade or business activity surveys. Private data providers such as Barbour ABI, Experian, Glenigan and Markit provide activity indicators. Government departments, the Bank of England and ONS provide large amounts of data that relates to or helps to inform on construction activity, from fixed capital formation, business demography, investment activity, employment and unemployment, pay, vacancies, redundancies, homes started and completed, planning and materials prices, stocks deliveries and imports. One can turn also to real estate data sources to provide indications of changing levels of activity within construction.
2 www.markit.com.
3 Information and guidance kindly provided by Frances Pottier, Senior Statistician and Deputy Head of Profession, Department for Business, Innovation and Skills.
4 In July 2016, the Department for Business, Innovation and Skills (BIS) and the Department of Energy and Climate Change (DECC) were merged to form the Department for Business, Energy and Industrial Strategy (BEIS). BIS was the sponsoring department for construction, this responsibility now falls within BEIS.
5 www.constructionproducts.org.uk.
6 www.experian.co.uk/economics/economic-forecasts/uk-construction-forecast.html.
7 "Companies House incorporates and dissolves limited companies, registers the information companies are legally required to supply, and makes that information available to the public. Companies House is an executive agency, sponsored by the Department for Business, Energy & Industrial Strategy". www.gov.uk/government/organisations/companies-house.
8 "The Statistical Classification of Economic Activities in the European Community, commonly referred to as NACE (for the French term *nomenclature statistique des activités économiques dans la Communauté européenne*), is the industry standard classification system used in the European Union". https://en.wikipedia.org/wiki/Statistical_Classification_of_Economic_Activities_in_the_European_Community.

References and further reading

Aldrick, P. (2011) ONS construction error moves markets. *The Telegraph*, 12 August. www.telegraph.co.uk/finance/newsbysector/constructionandproperty/8698951/ONS-construction-error-moves-markets.html.

Atkinson, A. (2005) *The Atkinson Review: Final Report. Measurement of Government Output and Productivity for the National Accounts* (Basingstoke: Palgrave Macmillan).

Barker, K. and Ridgeway, A. (2014) *National Statistics Quality Review: National Accounts and Balance of Payments*, Office for National Statistics. www.ons.gov.uk/ons/guide-method/method-quality/quality/quality-reviews/list-of-current-national-statistics-quality-reviews/nsqr-series–2–report-no–2–review-of-national-accounts-and-balance-of-payments.pdf.

BCIS (2017) www.rics.org/uk/knowledge/bcis.

Crook, T. and Sharp, G. (2010) Development of Construction Statistics. *Economic & Labour Market Review*, 4 (3). webarchive.nationalarchives.gov.uk/20160105160709/www.ons.gov.uk/ons/rel/elmr/economic-and-labour-market-review/no–3–march-2010/index.html.

Davies, J., Crook, T. and Greenaway, M. (n.d.) *Reviewing the Sample Design and Estimate Methodology for Output in the Construction Industry*, Office for National Statistics. www.ons.gov.uk/ons/guide-method/method-quality/ons-statistical-continuous-improvement/reviewing-the-sample-design-and-estimation-methodology-for-output-in-the-construction-industry/index.html.

Davies, K. and John, M. (2015) *Change to the Treatment of Outliers in Output in the Construction Industry*, Office for National Statistics. www.ons.gov.uk/ons/guide-method/method-quality/specific/business-and-energy/output-in-the-construction-industry/impact-of-improvement-to-outlier-treatment-for-output-in-the-construction-industry.pdf.

The Economist (2016) How to measure prosperity. *The Economist*, 30 April. www.economist.com/news/leaders/21697834-gdp-bad-gauge-material-well-being-time-fresh-approach-how-measure-prosperity.

Gov.UK (n.d.) *Live Tables on House Building: New Build Dwellings*. www.gov.uk/government/statistical-data-sets/live-tables-on-house-building.

Gov.UK (2014) *Construction Price and Cost Indices: New Methodology*, Department of Business, Innovation and Skills. www.gov.uk/government/publications/construction-price-and-cost-indices-new-methodology.

Gov.UK (2015) *Construction Price and Cost Indices*, Department of Business, Innovation and Skills. www.gov.uk/government/collections/price-and-cost-indices.

Gov.UK (2016) *Press Notice: 'Take economic statistics back to the future', Says Charlie Bean*. www.gov.uk/government/publications/independent-review-of-uk-economic-statistics-final-report.

Green, B. (2010) Is an inaccurate measure of foreign workers messing up the construction data? *Brickonomics: Building*, 21 February. http://brickonomics.building.co.uk/2010/02/is-an-inaccurate-measure-of-foreign-workers-messing-up-the-construction-data/.

Green, B. (2011) Why the lag in construction output data may be more important than the ONS seems to suggest. *Brickonomics: Building*, 16 December. http://brickonomics.building.co.uk/2011/12/why-the-lag-in-construction-output-data-is-more-important-than-the-ons-suggests/.

Green, B. (2015) The curious incident of the £2 billion boost to the UK's official annual construction output. *Brickonomics: Building*. 18 October. http://brickonomics.building.co.uk/2015/10/the-curious-incident-of-the-2-billion-boost-to-the-uks-official-annual-construction-output/.

Green, B. (2016) *Productivity in Construction: Creating a Framework for the Industry to Thrive*, Chartered Institute of Building. http://policy.ciob.org/research/productivity-construction-creating-framework-industry-thrive.

Hansard (1940) *Ministry of Works and Buildings*. hansard.millbanksystems.com/commons/1940/oct/24/ministry-of-works-and-buildings.

Kollewe, J. (2011) ONS admits it was wrong – there was no surprise spurt for UK construction. *The Guardian*, 13 August. www.theguardian.com/business/2011/aug/12/uk-construction-output-stronger-ons.

Laney, D. (2001) 3D data management: Controlling data volume, velocity and variety. In: *Application Delivery Strategies*, METAGroup. https://blogs.gartner.com/doug-laney/files/2012/01/ad949-3D-Data-Management-Controlling-Data-Volume-Velocity-and-Variety.pdf.

Lewis, D. (2007) Winsorisation for estimates of change and outstanding issues with the implementation of Winsorisation for level estimates. *13th Meeting of the National Statistics Methodology Advisory Committee*, Office for National Statistics (UK). www.ons.gov.uk/ons/guide-method/method-quality/advisory-committee/2005-2007/thirteenth-meeting/winsorisation-for-estimates-of-change.pdf.

ONS (n.d.[a]) *Business Investment in the UK Statistical Bulletins*, Office for National Statistics. www.ons.gov.uk/economy/grossdomesticproductgdp/bulletins/businessinvestment/previousReleases.

ONS (n.d.[b]) *Income and Wealth*, Office for National Statistics (UK). www.ons.gov.uk/peoplepopulationandcommunity/personalandhouseholdfinances/incomeandwealth.

ONS (n.d.[c]) *Construction Industry*, Office for National Statistics (UK). www.ons.gov.uk/businessindustryandtrade/constructionindustry.

ONS (n.d.[d]) *Questions and Answers Regarding the ONS Construction Output Survey*, Office for National Statistics (UK). www.ons.gov.uk/ons/rel/construction/output-in-the-construction-industry/june-and-q2-2011/questions-and-answers-regarding-the-ons-construction-output-survey.pdf.

ONS (n.d.[e]) *Inter-Departmental Business Register (IDBR)*, Office for National Statistics (UK). www.ons.gov.uk/ons/about-ons/products-and-services/idbr/index.html.

ONS (2007) *UK Standard Industrial Classification of Economic Activities 2007 (UK SIC 2007)*, Office for National Statistics (UK). www.ons.gov.uk/ons/guide-method/classifications/current-standard-classifications/standard-industrial-classification/index.html.

ONS (2010) *Output in the Construction Industry, March and Q1 2010*, Office for National Statistics (UK). www.ons.gov.uk/ons/rel/construction/output-in-the-construction-industry/q1-2010/index.html.

ONS (2011) *Output in the Construction Industry, Investigation into Lagged Responses*, Office for National Statistics (UK). http://webarchive.nationalarchives.gov.uk/20160105160709/www.ons.gov.uk/ons/rel/construction/output-in-the-construction-industry/investigation-into-lagged-response/index.html.

ONS (2014a) Output in the construction industry, October 2014 and New Orders Q3 2014. *Statistical Bulletin*, Office for National Statistics (UK). www.ons.gov.uk/ons/dcp171778_388276.pdf.

ONS (2014b) *National Accounts Articles – Impact of ESA95 Changes on Current Price GDP Estimates*, Office for National Statistics (UK). www.ons.gov.uk/ons/dcp171766_365274.pdf.

ONS (2014c) *New Orders in Construction QMI*, Office for National Statistics (UK). www.ons.gov.uk/businessindustryandtrade/constructionindustry/qmis/newordersinconstructionqmi.

ONS (2015a) *Construction Statistics*, Office for National Statistics (UK). www.ons.gov.uk/businessindustryandtrade/constructionindustry/articles/constructionstatistics/2015-08-26.

ONS (2015b) *Interim Solution for Construction Output Price Indices, Quarter 1 (January to March) 2015*, Office for National Statistics (UK). http://webarchive.nationalarchives.gov.uk/20160105160709/www.ons.gov.uk/ons/dcp171766_406521.pdf.

ONS (2015c) *ONS Census Transformation Programme: Administrative Data Update*, Office for National Statistics (UK). www.ons.gov.uk/ons/guide-method/census/2021-census/progress-and-development/research-projects/beyond-2011-research-and-design/research-outputs/administrative-data-update.pdf.

ONS (2016) *UK Non-Financial Business Economy (Annual Business Survey): Sections A-S*, Office for National Statistics (UK). www.ons.gov.uk/businessindustryandtrade/business/businessservices/datasets/uknonfinancialbusinesseconomyannualbusinesssurveysectionsas.

Pearce, D. (2003) *The Social and Economic Value of Construction*, Construction Industry Research and Innovation Strategy Panel (nCRISP). www.ccinw.com/images/publications/ncrisp%20the%20social%20and%20economic%20value%20of%20construction.pdf.

UK Statistics Authority (2011a) *Review of an Error in the Published Estimate of Output in Construction*, UK Statistics Authority. www.statisticsauthority.gov.uk/news/review-of-an-error-in-the-published-estimates-of-output-in-construction/.

UK Statistics Authority (2011b) *Assessment of Compliance with the Code of Practice for Official Statistics: Construction Price and Cost Indices*, UK Statistics Authority. www.statisticsauthority.gov.uk/wp-content/uploads/2015/12/images-assessment-report-95-construction-price-and-cost-indices_tcm97–38289.pdf.

US Census Bureau (n.d.) *The X-!#ARIMA-SEATS Seasonal Adjustment Program*, United Stated Census Bureau. www.census.gov/srd/www/x13as.

WTO (2008) The conceptual framework for tourism statistics – International Recommendations for Tourism Statistics 2008 (IRTS 2008). *Statistics and Tourism Satellite Account*, World Tourism Organization. http://statistics.unwto.org/content/irts2008.

Yu, M. and Ive, G. (2015) A review of construction cost and price indices in Britain. In: Best, R. and Meikle, J. (eds.) *Measuring Construction: Prices, Output and Productivity* (London: Routledge).

Editorial comment

It is predicted that cash purchases in the UK will fall to just 21% of total sales by 2026 (Lyons *et al.* 2018). In spite of a clear trend away from cash to digital payments, the so-called cash economy (also variously referred as the shadow economy, black economy, cash-in-hand economy, gig economy and hidden economy – while often used interchangeably, these terms do have some variations) is still evident in most countries. In 2017 it was estimated that 1.1 million people in the UK were working in the 'gig economy', with 18% of them in plumbing, building and other skilled manual work (BBC 2017). It is also certain that many others receive at least occasional cash-in-hand payments, particularly for small trade jobs such as installing a light fitting or fixing a leaky tap.

The total cash economy in Australia is estimated to be worth as much as AUD 50 billion per annum (Khadem 2018); however, that includes a wide range of components including under-reporting of income, identity fraud, money laundering and illegal drug profits, as well as unreported cash payments for work such as the small trade jobs noted above. By its nature it is very difficult to make any sort of reliable estimate of just how big the shadow economy is in any country, and it is equally difficult to estimate how much of the unreported activity is related to construction.

What is sometimes referred to as informal construction may account for a very large proportion of total construction in some countries where only construction done by contractors is recorded and a lot of work is done by building owners, perhaps with assistance from family and other people around them who are either unpaid or paid in kind. In all cases the result is some degree of under-reporting of total construction activity and value. In countries such the UK and Australia, governments are concerned not only about illegal activities, such as drug trafficking, and the economic activity associated with such activities, but they also see a significant loss of tax revenue related to unrecorded cash payments for work done across many fields.

Measuring the shadow economy and its potential impact on measures of output and productivity requires further investigation as researchers and government agencies seek to improve and refine their methods. In this chapter the authors discuss key aspects of the impact of the shadow economy, particularly in the context of construction statistics and national accounts. They also describe a number of attempts that have been made to estimate the value of unrecorded work in a number of countries and consider the usefulness and applicability of the various methods.

5 Measuring construction industry activity and productivity

The impact of the shadow economy

Will Chancellor, Malcolm Abbott and Chris Carson

Introduction

The shadow economy[1] is often not considered in productivity research, as its impact on productivity estimation is seen in most cases as being negligible. In general, this is a realistic assumption at the national level, but in the case of some industries there is perceived to be a disproportionately sized and significant shadow economy. Most studies, for instance, estimate the construction industry as possessing the greatest shadow economy compared to normal production (Schneider 2013; ABS 2013a). Shadow economic activity can be defined as being those 'activities that are productive and legal but are deliberately concealed from the public authorities to avoid payment of taxes or complying with regulations' (OECD 2002: 13).

Within the context of the construction industry this activity usually takes the form of short-term and labour intensive 'cash-in-hand' work. Income can also include proceeds of barter trade or electronic funds; however, it is generally accepted that most transactions occurring in the shadow economy are in cash. Income is then concealed from reporting records, typically for the purposes of avoiding taxation expenses. By operating in the shadow economy, the participant may also be avoiding regulatory responsibilities such as safety and quality compliance, with the motive to further increase profit margins. These taxation and regulatory shortcuts are detrimental to the legitimate businesses operating within a competitive market, to the quality of products and services purchased by consumers and to taxation revenue.

Within the subdivisions of the construction industry, it is expected that most shadow activity occurs within residential construction services as opposed to heavy civil engineering and large-scale building construction. The small scale of residential construction services such as building a fence, installing a pipe or repairing a plaster wall are likely to pose lower risk in concealing monetary gain from authorities. Conversely, the construction of a freeway bridge requires large-scale investment and media attention, meaning that it would be difficult to conceal from authorities and highly unlikely to be completed by shadow economy operators.

Fichtenbaum (1989), when he undertook investigations into the relationship between the shadow economy in the United States and Canada and the slow growth in productivity in the 1970s and 1980s, came to the conclusion that it was exaggerated because of the unreported growth of the shadow economy. As with many shadow economy studies, however, the focus in Fichtenbaum (1989) was the overall economy rather than the construction industry specifically. Similarly, Meon *et al.* (2011), in investigating the relationship between national shadow economies and technical efficiency, found that a failure to make adjustments for the shadow economy led to errors in measured efficiency. In addition, Bhattacharyya (1999), in considering the impact of the shadow economy, found that the perceptions of standard economic relations may be distorted if the shadow economy is not taken into account, and these distorted perceptions may have an impact on policy analysis.

In the case of the construction industry there has been, in recent years, a rise in interest in studying the industry's efficiency and productivity. In part this interest has manifested itself in the use of a number of statistical techniques by researchers attempting to gauge the efficiency and productivity of the industry. However, even if estimates of shadow activity are included, there is some degree of uncertainty about how significant this activity is, and that these estimates can be affected by decisions about which alternative method of estimation is used.

The purpose of this chapter is, therefore, to summarise the key findings determined from the past research into measurement of construction activity, focusing on the estimation problems associated with the construction industry shadow economy (including whether an increase in the shadow economy might be making increased supply) and suggest some of the ways that these problems might be addressed. In looking at how shadow activity is incorporated into productivity estimates, the issue of the impact of the alternative methods (which can affect the size of estimated shadow activity) will be considered.

How big is the shadow economy?

One of the few consistencies in shadow economy research is that the construction industry tends to stand out as being more actively involved than other industries in developed economies. This was well documented in the Australian Bureau of Statistics information paper (ABS 2013a), where industry level shadow economy production was estimated for the 2010–2011 financial year. Through the use of a variation of the national accounting method, the Australian Bureau of Statistics (ABS) applied audit information from the Australian Taxation Office to adjust their shadow economy estimates. These adjustments were then applied to several components of gross domestic product (GDP) including gross mixed income, gross operating surplus, gross output, total intermediate use and household final consumption expenditure. There, findings indicated that the construction industry shadow economy accounted for 10.2% of construction industry gross value added (GVA), or approximately $10 billion. This was considerably higher than the industry average of 1.6% revealed in their study. The industry displaying the next largest shadow economy was accommodation and food services, with 2.6% of its industry GVA being undertaken in the shadow economy, equating to approximately $850 million for the 2010–2011 financial year. The results published by the ABS could be considered as conservative, with the average percentage aligning to previous conclusions in their 2003 study, which also regarded the shadow economy as negligible. While ABS (2003) does not specifically focus on individual industries, it presents a similar aggregate estimate, that is that the Australian shadow economy is highly unlikely to exceed 2% of GDP. The specific percentage of total shadow economic output in ABS (2003) was estimated as 1.6%, or AUD 20.7 billion.

A comparable study was undertaken by Chancellor and Abbott (2015), focusing specifically on the Australian construction industry shadow economy using the Multiple Indicator Multiple Cause approach. Despite using a completely different approach, they arrived at a similar shadow economy estimate for the Australian construction industry to that reported by the Australian Bureau of Statistics (ABS 2013a). They found that the construction shadow economy had been generally increasing relative to construction industry GVA. Reinhart *et al.* (2004) provide another, earlier Australian example; unlike the studies reported in ABS (2013a) or Chancellor and Abbott (2015), they used a direct survey approach to estimate the shadow economy. Their findings indicate engagement rates of between 5% and 15% in the shadow economy, with labour and trade type workers being the most active participants. Putnins and Sauka (2011) studied the shadow economies of Estonia, Latvia and Lithuania using the survey method. They found the construction industry to be actively involved in both Estonia and Lithuania, and to have the largest shadow economy of all industries in Latvia.

Schneider (2013) reviewed the shadow economy in Europe, noting that the construction industry has always made up a large proportion of total shadow economic activity. His estimates indicate that almost a third of construction industry GVA is occurring in the shadow economy. Williams *et al.* (2011) also examined the European construction industry shadow economy. Using data collected in the 2007 Eurobarometer survey on undeclared work, which interviewed 26,659 respondents in 27 European Union member states, they found 16% of undeclared jobs in Europe occurred in the construction industry. They also concluded that 18% of the construction labour force engaged in some form of undeclared activity in the 12 months prior to being surveyed.

With regard to the Danish economy, Schneider notes that the building and construction industry was also prominent in the shadow economy. Based on data from Hvidtfeldt *et al.* (2011: 5) for 2010, of the 32% of the people who had carried out undeclared work in the previous 12 months, 48% had carried out building and construction work. This information was obtained using a survey method. During the collection of data, it was also noted that the survey participants were tolerant or accepting of the shadow economy in general.

Different methods for shadow economy estimation

Attempting to estimate the size of the shadow economy within the construction industry presents a number of challenges. Fundamentally, the concealed nature of the shadow economy means that it is not possible to simply download a data-set from the national statistical office. The shadow economy needs to either be estimated based on known variables or be collected directly using a statistical instrument such as a survey. The various approaches have their own benefits and limitations and it is important to be aware of the advantages and disadvantages before attempting to apply any of these methods.

A variety of methods have been implemented in economic research; however, the most common are categorised as either direct or indirect methods. The most commonly used indirect methods are the consumption method, the emission method, the national accounting method, the currency ratio method and the Multiple Indicators Multiple Causes method. A range of variations to these indirect methods have been developed, often to include some new component in an attempt to improve the accuracy of the estimate. The following section will focus on the most commonly used methods for shadow economy estimation within the context of construction industry research.

Consumption method

The electricity consumption method is a popular choice for shadow economy estimation and has been used in important studies such as Kaufman and Kaliberda (1996). It is, however, less frequently used in construction industry–level studies. The fundamental theory behind the consumption method is the assumption that growth in electricity or some other type of energy consumption, such as diesel or petrol, relative to economic growth provides an indication of change in the shadow economy. Growth in energy consumption and growth of the observed economy, in theory, should coincide over time and a divergence in trends between these variables may indicate that the shadow economy is either increasing or decreasing. If, for example, electricity consumption is increasing at a faster rate than economic growth, the additional electricity consumption may indicate a growing shadow economy.

An advantage of the consumption method is its simplicity. The data required is usually basic and available, particularly for national level analysis. It is, however, something of a blunt instrument in shadow economy trend estimation, especially when focusing on a particular industry such as construction. This method also has some limitations. An immediate limitation in the context of the construction industry is that production activity in the industry is predominantly labour intensive, sometimes requiring little or no consumption of electricity or other energy resources. It is entirely possible for production to take place within the construction shadow economy that uses no electrical energy, such as hammering nails into a fence or digging a hole with a shovel. Conversely it is also possible to consume high amounts of electricity in activities such as welding metal or cutting timber. The point is that electricity consumption in the construction industry is not predictable; rather it can vary considerably based on the production activity being undertaken. The electricity consumption in other industries such as financial services is potentially more predictable, since the production activity requires more constant use of equipment such as computers, telephones and lighting. Therefore, when estimating for a specific industry, consideration should be given to the energy intensity of the target industry compared to other industries. These variations in consumption intensity tend to balance out over broader studies that focus on national economies rather than on specific industries.

An additional consideration with regard to construction industry–specific shadow economy estimation is that production activity occurring in the shadow economy is generally lower-skilled and highly labour-intensive work such as residential construction and various construction services such as plumbing or bricklaying. As noted earlier, it would be unlikely that heavy civil engineering construction would be occurring within the construction shadow economy due to the difficulties in concealing such large-scale activity. It could also be argued that the labour-intensive work occurring in the shadow economy relies less on the consumption of energy than heavy civil and building construction. Another concern is that a reduction in energy consumption compared to economic growth might be due to more efficient technology and processes that have developed over time. The consumption method, however, continues to be useful in the calculation of national level shadow economies and the construction-specific shadow economy.

An example of how the electricity consumption method could be applied for shadow economy estimation is presented in Equation 5.1. This methodology is based on Kyle *et al.* (2001) and assumes a constant GVA to electricity ratio with a base year of 1975. Equation 5.1 requires an input estimate (x) for the shadow economy in the base year. Since an accurate estimate is usually not available, a range of possible shadow economy values in the base year can be substituted. By alternating these values, it is possible to demonstrate that regardless of the base year estimate, the shadow economy trend remains the same. This is useful in predicting how the shadow economy has changed over time and a possible range for the shadow economy size for the present. To apply this method specifically to the construction industry, it is possible to substitute national GVA with construction industry value added. In doing so, however, the shadow economy estimate would be based on electricity consumption for the entire economy. Therefore, if this method were to be used to estimate the construction shadow economy, it may be more suitable to use it in conjunction with other methods in order to compare results and establish a more accurate estimate.

$$SE_{1975} = (1+x) * \left(\frac{Y_{base}}{E_{base}} \Big/ \frac{Y_{1975}}{E_{1975}} \right) - 1 \qquad \text{(Eq. 5.1)}$$

SE = shadow economy, Y = gross value added, E = electricity consumption

Another variation of the consumption method is the energy consumption method, one that is based on the same principles and methodology as the electricity consumption method, yet uses consumption of other resources including liquefied petroleum gas (LPG), petrol, diesel, kerosene, diesel oil, fuel oil, gas and electricity, rather than electricity only. Depending on data availability, this is considered to be a more complete shadow economy estimate by comparison to the electricity consumption method. Kyle *et al.* (2001) suggest that the inclusion of 'total energy' is an improvement on the electricity consumption method, particularly as a construction worker operating in the shadow economy may alternate between energy sources, from using the residential electricity supply to using a portable generator that consumes petroleum fuel. In addition, the consumption of energy from different sources could be used to identify shadow economy variations for different segments of the construction industry. Consumption of diesel could coincide with the use of heavy machinery in the civil engineering and building construction segments of the industry, whereas typical construction services, such as those likely to occur in the shadow economy, may use only electricity and petrol. Often, however, the possibility of observing these variations and making these assumptions will rely on availability of data.

With consideration being given to the consumption of energy inputs for shadow economy estimation, it is also worthwhile noting that the reverse is possible. The emission method compares pollution output, rather than energy input, to economic growth, with the theory being that increases in certain emissions such as carbon should coincide with increases in economic production. As with most shadow economy estimation techniques, the limitations need to be fully understood. After all, in its basic form this theory does not consider that economic production may have become more environmentally efficient over time and that certain industries or even specific processes emit considerably higher emissions than others. Another serious limitation, particularly within the context of the construction industry, is that it is difficult to collect industry specific pollution data; rather this data is typically more general, such as total megatons of carbon dioxide for an entire country. In some instances, firm-level emission data may be available to use as a sample of construction industry pollution.

National accounting method

The Organisation for Economic Co-operation and Development's (OECD 2002) *Measuring the Non-observed Economy* handbook provides instructions for estimating shadow economic activity using the national accounting method, particularly in relation to data confrontation and discrepancy analysis. Examples of the specific discrepancy methods outlined include comparing the actual amount of taxes collected with the theoretical amount of tax that should have been collected; comparison between personal income from tax returns and household income from the national accounts; the discrepancy between income-based and expenditure estimates of national income (also in O'Higgins 1989; MacAfee 1980); comparison of wages and employment from the supply and use tables which compares data from the employer against data from the employee; other micro-discrepancy analysis that compares data for individuals or samples of people or companies; and partial integration, which is a confrontation of two related data sources such as the production of construction materials compared to construction outputs. This method is used by national statistics offices as demonstrated by the example of the Australian Bureau of Statistics (ABS 2013a), which presents estimates of the shadow economy, extending the scope of the national accounting method to industry level shadow economy estimates.

Another national accounting-based method for shadow economy estimation described in OECD (2002) is referred to as sensitivity analysis or upper bound estimation. The likely maximum amount of shadow economic activity in existence is included and then a new GDP is calculated. The difference between the new GDP and the old GDP figures give an estimate of the maximum value of the shadow economy. An applied example for Canada in 1994 is given in OECD (2002), with an explanation provided for each adjustment. In this case, for each component of GDP an upper bound on shadow activity was computed and then compared with production levels in official estimates. The main areas computed were residential construction and final consumption expenditure of households, it being envisaged that shadow activity in imports and exports was quite small and on other expenditure components of GDP was quite negligible. In the case of residential construction housing starts, the average value of building permits and information from the householder repair and renovation survey were used to estimate shadow activity. In the case of residential expenditure, estimates were made of expenditure in 140 different categories with particular emphasis on tobacco, alcohol, rent, room and board, professional services, childcare, food, and domestic and household services (OECD 2002: 55–57). This method appears to be designed to estimate shadow economic activity for an entire economy rather than a specific industry. However, provided that all elements of industry-level national accounting data are available, then an industry-level estimation for shadow economic activity is possible.

Multiple Indicators Multiple Causes (MIMIC)

One method of shadow economy estimation gaining in popularity in recent years is the Multiple Indicators Multiple Causes (MIMIC) model referred to earlier (Schneider *et al.* 2010; Schneider and Bajada 2003; Schneider 2006; Meon *et al.* 2011; Chancellor and Abbott 2015). This method relies on looking at the factors that are regarded as causing the shadow economy and factors which indicate the existence of the shadow economy, to derive a latent or 'unknown' shadow economy variable. Depending on data availability, it is possible to produce both economy-wide and industry-specific estimates. The actual estimation itself uses a form of structural equation modelling that can be undertaken in some statistical software programs such as SPSS AMOS. Variations in the shadow economy estimate will depend largely on the causal and indicator variables chosen, as well as the extensiveness of the data. Causal variables are typically factors such as taxation burden and unemployment rate. Indicator variables can include factors such as industry GVA or cash in the economy. The shadow economy estimate will depend entirely on the causal and indicator variables chosen, therefore it is important that these variables are defined carefully.

As with most shadow economy methods, the MIMIC approach is often applied at the national economy level rather than for specific industries such as construction. To apply this method to the construction industry is potentially more challenging due to the unavailability of causal and indicator variables at this lower level. It is possible, however, to use the MIMIC method to produce construction industry shadow economy estimates (see Chancellor and Abbott 2015). Ideally all causal and indicator variables will specifically relate to the construction industry; however, it is also possible to include more general variables such as changes in taxation burden over time and cash in the economy. Once calculated, the latent variable or shadow economy variable can be converted into meaningful figures.

Currency ratio methods

This method assumes that the amount of cash in an economy over time is an indication of the levels of shadow economy activity. The general thought is that to conceal transactions in the shadow economy, participants will conduct all business in cash only, which of course is not always the case – many transactions are expected to occur using barter trade or, more commonly, other electronic transfer methods. The currency method is outlined in Feige (1980, 1986). There are a number of variations of this method; for example, Cebula and Feige (2011) attempted to address the shift towards electronic funds transfers by applying a relaxed assumption to deposit holdings and the currency ratio. This method is more often applied to entire economies rather than specific industries such as construction. In an attempt to use this method to estimate cash-in-hand activity compared to economic growth in the construction industry, it would be possible to substitute total economic growth data with industry specific growth data such as construction industry GVA. The obvious limitation is that data for cash in the economy does not relate specifically to construction, but rather to the entire national economy. Either way, it might provide some indication or a possible range for the construction shadow economy in a similar way to the consumption method.

Direct method – survey approach

The direct method or direct collection approach is another means by which a shadow economy can be estimated. For this method, data is collected directly from respondents, usually with the assistance of a survey form or other survey instrument. As with all shadow economy methods, the survey approach has its own advantages and disadvantages. The cost to develop and implement a survey is potentially prohibitive, as is the selection of an unbiased survey sample. Another challenge is that some respondents may be fearful of disclosing information that may be considered as an admission of taxation or regulatory evasion. For this reason, the assumption could be made that survey results provide a deflated estimate of shadow economic activity. The main benefit of this method is that data obtained directly from the shadow economy is being used which, provided the survey and sampling methodology is robust, may result in the most accurate estimation available. Other modelling and indirect techniques risk providing a vague indication of shadow economic activity, particularly at lower industry levels.

There are few examples of direct collection methods being used to specifically estimate construction shadow economic activity, rather these studies tend to be broader economy-wide studies whereby specific industry details are researched in aggregated and disaggregated format. An extensive study, the 2007 Eurobarometer survey, was conducted in Europe with data gathered from 26,659 respondents in 27 European Union member states; one of its purposes was to estimate the shadow economies. As with the findings in ABS (2013a), the results in the 2007 Eurobarometer suggest that the construction industry makes up a significant proportion of the shadow economy. Another example of the direct collection method is found in Reinhart *et al.* (2004); they surveyed several broad occupations and concluded that labourers and trade employees were the most active in the Australian shadow economy. As with the findings in Williams *et al.* (2011), Reinhart *et al.* (2004) also noted that a significant proportion of shadow economic activity was undertaken by people who were also employed in the official economy. In the survey-based Putnins and Sauka (2011) study mentioned earlier, the researchers, clearly aware of the reporting challenges in collecting sensitive and potentially incriminating data, began their questionnaire with non-sensitive questions about external factors before moving to the more sensitive shadow economy questions. They conducted their questionnaire over the phone, interviewing 500 participants in Estonia, 591 in Latvia and 536 in Lithuania.

Adjusting construction productivity to account for the shadow economy

Previous studies that have attempted to estimate the shadow economy within the construction industry have produced similar conclusions. The construction industry in most developed economies is likely to be one of the most actively involved in shadow economic activity compared to other industries (OECD 2002: 76). Both cash-in-hand paid to builders and a lot of home renovation work that is not recorded make the sector's shadow bigger. This is, however, more likely to relate to residential building rather than other sectors of the industry. Consideration should therefore be given to the effect the shadow economy may have on other economic statistics such as productivity. Unreported data has the potential to influence productivity estimation as discussed by Fichtenbaum (1989). A theory discussed in Chancellor and Abbott (2015) is that capital inputs are likely to be reported for shadow economy production, and rather it is the labour input and production output that is potentially concealed by operating in the shadow economy. Construction capital inputs are expected to be purchased through normal channels regardless of the production activity occurring in the observed or unobserved economy. Chancellor and Abbott (2015) note that the inclusion of capital inputs and the partial inclusion of labour input and production output are expected to put downward pressure on official estimates of construction productivity growth.

Chancellor and Abbott (2015) also propose that the construction shadow economy operates more freely, without the burden of safety or regulatory standards, and can therefore expect higher levels of technical efficiency and subsequent productivity growth compared to the official economy. A similar discussion is noted in Williams *et al.* (2011); they comment that the European construction industry shadow economy attracts investment due to its ability to overcome limitations in the official economy resulting from regulatory burden and cost constraints.

There are a limited number of examples in which construction industry productivity has actually been adjusted to account for the shadow economy. The greater challenge is obtaining an industry level estimate. Provided estimates can be determined, it is then technically possible to use this data for the purpose of adjusting a construction productivity time series to account for the shadow economy. This adjustment could be made to productivity inputs or outputs prior to productivity estimation or to the final productivity estimate itself. This field of research is fairly new, with only one study identified in which an attempt has been made to adjust productivity based on shadow economic activity (Chancellor and Abbott 2015). In this study, construction industry productivity is estimated over time using Färe-Primont Data Envelopment Analysis. The productivity estimates are then adjusted to account for the construction industry shadow economy by including a non-discretionary adjustment variable quality adjustment as described in Coelli *et al.* (2005: 192). There is considerable scope to explore other methods of productivity adjustment based on shadow economy estimates.

Conclusion

As the examples have demonstrated, the construction industry is likely to be the most active participant in the shadow economy compared to other industries in the developed world. In an Australian context, billions of construction dollars are going unobserved and untaxed. From a productivity perspective, however, the unobserved statistical inputs and outputs should not be ignored, as demonstrated by Chancellor and Abbott (2015). The shadow economy has the potential to distort construction productivity, given the high amounts of unobserved inputs and outputs.

There is no simple solution to the problem of how to adjust productivity in a way that accounts for the shadow economy. Rather it is a matter of acknowledging that the shadow economy may have distorted the productivity estimates and to test shadow economy estimation methods based on available data. Although potentially time-consuming, applying several methods in conjunction appears to be the most suitable way of obtaining more robust measurements of shadow economic activity in the construction industry. As this field of research develops, new methods that account for construction shadow economic activity may be developed and refined. Another option is to use existing research that has obtained an estimate for the construction industry shadow economy, such as ABS (2013a), Williams *et al.* (2011) or Chancellor and Abbott (2015). These estimates could then be applied in various ways to adjust productivity estimates.

A number of past studies identify the construction industry as being a significant contributor to the overall shadow economy. Few studies, however, drill further into the problem by focusing on the construction industry shadow economy or identifying the effect of the shadow economy on other economic indicators. Future research opportunities exist in further estimating industry specific shadow economic activity, either through survey or indirect approaches.

Note

1 Sometimes referred to as the black or underground economy. The term shadow economy is preferred here as it is true that some shadow construction activity could be illegal, the black economy definitions are usually used when measuring obviously illegal activity such as trade in illegal drugs (see, for example, ABS 2013b).

References and further reading

ABS (2003) *Australian Economic Indicators, Oct 2003: Feature Article – The Underground Economy and Australia's GDP*, Cat. No. 1350.0 (Canberra: Australian Bureau of Statistics).

ABS (2013a) *Information Paper: The Non-Observed Economy and Australia's GDP, 2012*, Cat. No. 5204.0.55.008 (Canberra: Australian Bureau of Statistics).

ABS (2013b) *Measuring Illegal Production* (Canberra: Australian Bureau of Statistics). www.abs.gov.au/ausstats/abs@.nsf/Latestproducts/5204.0.55.008Main%20Features620 12?opendocument&tabname=Summary&prodno=5204.0.55.008&issue=2012&num= &view=.

Bajada, C. (1999) Estimates of the underground economy in Australia. *Economic Record*, **75**, 369–384.

BBC (2017) Taylor review: All work in UK economy should be fair. *BBC News*. www.bbc.com/news/business-40561807.

Bhattacharyya, D. K. (1999) On the economic rationale of estimating the hidden economy. *The Economic Journal*, **109**, 348–359.

Cebula, R. and Feige, E. (2011) America's unreported economy: Measuring the size, growth and determinants of income tax evasion in the U.S. *Crime, Law and Social Change*, **57** (3), 265–285.

Chancellor, W. and Abbott, M. (2015) The Australian construction industry: Is the shadow economy distorting productivity? *Construction Management and Economics*, **33** (3), 176–186.

Coelli, T., Prasada Rao, D. S., O'Donnell, C. and Battese, G. E. (2005) *An Introduction to Efficiency and Productivity Analysis*, 2nd ed. (New York: Springer Science Business Media, LLC).

Feige, E. L. (1980) *A New Perspective on Macroeconomic Phenomena: The Theory and Measurement of the Unobserved Sector of the United States: Causes, Consequences, Implications* (Denver, CO: Paper presented at the American Economics Association Meeting).

Feige, E. L. (1986) A re-examination of the underground economy in the United States. *IMF Staff Papers*, **33** (4), 768–778.

Fichtenbaum, R. (1989) The productivity slowdown and the underground economy. *Quarterly Journal of Business and Economics*, **28** (3), 78–88.

Frey, B. and Weck, H. (1983) Estimating the shadow economy: A 'naive' approach. *Oxford Economic Papers*, **35** (1), 23–44.

Hvidtfeldt, C., Jensen, B. and Larsen, C. (2011) *Undeclared Work and the Danes*, University Press of Southern Denmark, June 2010, English summary reported in: Rockwool Foundation Research Unit, News, March 2011, Copenhagen, Denmark.

Kaufmann, D. and Kaliberda, A. (1996) Integrating the unofficial economy into the dynamics of post socialist economies: A framework of analyses and evidence. In: Kaminski, B. (ed.) *Economic Transition in Russia and the New States of Eurasia* (London: M.E. Sharpe), 81–120.

Khadem, N. (2018) Is the cash economy really worth $50b and who are we fighting? *Sydney Morning Herald*, 26 June. www.smh.com.au/money/tax/is-the-cash-economy-really-worth-50b-and-who-are-we-really-fighting-20180626-p4znqp.html.

Kyle, S., Warner, A., Dimitrov, L., Krustev, R., Alexandrovna, S. and Stanchev, K. (2001) The shadow economy in Bulgaria. Harvard University, Agency for Economic Analysis and Forecasting and Institute for Market Economics.

Lyons, K., Jones, R. and Collinson, P. (2018) Revealed: Cash eclipsed as Britain turns to digital payments. *The Guardian*, Tuesday 20 February. www.theguardian.com/money/2018/feb/19/peak-cash-over-uk-rise-of-debit-cards-unbanked-contactless-payments.

MacAfee, K. (1980) A glimpse of the hidden economy in the national accounts. *Economic Trends*, **136**, February, 81–87.

Meon, P., Schneider, F. and Weill, L. (2011) Does taking the shadow economy into account matter when measuring aggregate efficiency? *Applied Economics*, **43** (18), 2303–2311.

O'Donnell, C. J. (2011) *A Program for Decomposing Productivity Index Numbers* (Brisbane: Centre for Efficiency and Productivity Analysis, The University of Queensland).

OECD (2002) *Measuring the Non-Observed Economy – A Handbook*. Organisation for Economic Co-operation and Development (Paris: OECD Publications Service).

O'Higgins, M. (1989) *Measuring the Hidden Economy: A Review of Evidence and Methodologies* (London: Outer Circle Policy Unit).

Putnins, T. J. and Sauka, A. (2011) Size and determinants of shadow economies in the Baltic States. *Baltic Journal of Economics*, **11** (2), 5–25.

Reinhart, M., Job, J. and Braithwaite, V. (2004) *Untaxed Cash Work: Feeding Mouths, Lining Wallets, Regulatory Institutions Network* (Canberra: Australian National University).

Schneider, F. (2006) Shadow economies of 145 countries all over the world: What do we know? Discussion paper, Department of Economics, University of Linz, Linz, Austria.

Schneider, F. (2013) *The Shadow Economy in Europe, 2013* (Linz: Johannes Kepler Universität).

Schneider, F. (2014) *The Shadow Economy and Shadow Labour Force: A Survey of Recent Developments*, IZA Discussion Paper No. 8278 (Bonn: Institute for the Study of Labor).

Schneider F. and Bajada C. (2003) The size and development of the shadow economies in the Asia-Pacific, Economics Working Papers 2003–01, Department of Economics, Johannes Kepler University, Linz, Austria.

Schneider, F., Buehn, A. and Montenegro, C. (2010) *Shadow Economies All Over the World; New Estimates for 162 Countries from 1999 to 2007*, WPS5356 (Washington: The World Bank Development and Research Group Poverty and Inequality Team and Europe and Central Asia Region Human Development Economics Unit). http://dx.doi.org/10.1596/1813-9450-5356.

Shephard, R. W. (1953) *Cost and Production Functions* (Princeton: Princeton University Press).

Shephard, R. W. (1970) *The Theory of Cost and Production Functions* (Princeton: Princeton University Press).

Williams, C., Nadin, S. and Windebank, J. (2011) Undeclared work in the European construction industry: Evidence from a 2007 Eurobarometer survey. *Construction Management and Economics*, **29** (8), 853–867.

Editorial comment

A key concern in any methodology used for the measurement of construction productivity is the definition of *construction output* and how that may be measured. In manufacturing, output is often reasonably easy to identify: a unit of output may be a motor vehicle, a television set or even the mythical widget so often mentioned in economics textbooks. Even using that sort of measure of output is not as straightforward as it may appear, as a small sedan is hardly the same as a large SUV or a luxury limousine and a portable television is not the same as a large, curved-screen smart TV. Each requires different inputs (different in type, quantity and quality), and the monetary value of the inputs for each will vary considerably as will the value of each unit of output.

In construction the problem is even more complex. Even standard house designs are routinely customised to suit individual client requirements and adjusted to fit onto different sites, while bespoke house designs range from compact inner-city houses to large mansions with home theatre, gymnasium, indoor swimming pool, garaging for multiple vehicles and more. Non-residential buildings present further problems. Measuring and/or comparing productivity across different cultures, regulatory frameworks, material and labour supply chains and climatic zones just adds to the difficulties.

In this chapter the authors discuss a variety of approaches to the problem of defining and measuring output and categorise many studies in which researchers around the world have tried to address productivity measurement. They focus particularly on measures of output and price changes over time as key problem areas of construction productivity measurement, but also identify a number of other productivity-related issues. These include the fact that off-site production of components for construction may lead to resulting productivity improvements being allocated to manufacturing industries rather than construction; that the prevalence of shadow construction activity in most countries will often distort measurements of productivity; and that accounting for quality improvements over time can have an impact on productivity measures. These all merit the attention of construction researchers; the last mentioned is discussed in Chapter 5. As part of the discussion the authors also consider the validity and reliability of some construction statistics, a theme that is also explored by others elsewhere in this book.

6 Productivity and levels of output in the construction industry

Will Chancellor, Malcolm Abbott and Chris Carson

Introduction

One of the problems with determining levels of productivity in the construction industry has been the difficulties associated with determining levels of output. Using physical measures is problematic because of the heterogeneity of construction, and the traditional financial measure of value added is also problematic as the industry increasingly moves to off-site fabrication. As work moves off-site, work that is done on-site becomes increasingly unskilled and apparently becomes less productive. In addition, this off-site work output may have shifted away from the construction industry and into part of the manufacturing industry, resulting in further measurement and estimation challenges. The purpose of this chapter, therefore, is to raise some of the issues involved in using various indicators of output in productivity analysis.

Productivity measures

In determining the levels and change of productivity of an industry it is crucial to be able to understand the nature of the data used. At its most basic level, productivity is the ratio of output created by an industry (or firm or country) divided by the inputs used to produce it. If this ratio rises, then it indicates that more is being produced from a given level of inputs and that therefore productivity is improving. The accuracy of the productivity measures therefore depends on the data that is used to indicate the outputs and inputs and the manner in which they are compared to each other.

The way construction industry productivity has been analyzed over the years, and the way in which outputs and inputs have been compared to each other, has been influenced by the manner in which productivity analysis more generally has developed. In undertaking productivity analysis, researchers have used a range of productivity and efficiency measurement techniques. In the past, productivity changes over time were first measured using an 'index' approach. This approach involves the construction of index numbers which can be used to indicate the partial or total factor productivity of an industry. Partial productivity measures generally relate a firm's (or industry's) output to a single input factor – for example, in the construction industry the volume of construction activity per employee is often used as a labour-based partial productivity measure (see examples of this approach used in Table 6.1). Total factor productivity measures are generally the ratio of a total aggregate output quantity index to a total aggregate input quantity index (and total factor productivity growth is then the difference between the growth of the output and input quantity indices). Total factor productivity indices were developed from the early 1950s, especially by the National Bureau of Economic Research in the United States, a pioneering institution in the development of productivity analysis. One of its employees in particular, John W. Kendrick, published extensively using these techniques for a number of years (Kendrick 1956a, 1956b, 1961, 1973). Amongst those industries examined by the Bureau and Kendrick was the construction industry, and these were the first nationwide construction industry productivity estimates. Other researchers also began to undertake similar studies of this sort from the 1950s onwards (see Table 6.1). Over the years, studies have also been conducted in a range of other countries besides the United States using the index approach. Besides the index approach, econometric measures have also been used which apply estimations of cost or production functions. The estimated functions can then be used to identify changes in productivity or productive efficiency. In addition, data envelopment analysis (DEA), which is a linear programming technique, has been used to estimate productivity levels in the construction industry. Probably the first example of the DEA Malmquist approach being used to determine the change in productivity in the construction industry over time was undertaken by Färe, Grosskopf and Margaritis (1996) for the New Zealand construction industry. This work was undertaken as part of a broad economy-wide study of productivity change in New Zealand, where the work on the construction industry was just one of a number of industry-level studies undertaken. Since this work was done, other studies have been completed in a number of other countries (Table 6.1). As well as depicting changes in productivity over time in the construction industry, there have also been examples of DEA being used to benchmark firms in the industry. It is notable, however, that in the selection of inputs and outputs using DEA as a benchmarking tool, the choices are more varied than is the case in time series studies.

In addition to the three approaches used, research on construction industry productivity has also been concerned with site-level labour productivity, which has a more direct relevance to industry management (see, for instance, Ganesan 1984; Lowe 1987; Maloney 1983; Allen 1985; Thomas and Sakarcan 1994). This is also due to the fact that in the construction industry no single construction project is exactly the same as any other, and no construction company is strictly speaking the same in composition as any other. For that reason, a number of the studies in this area fall back on using aggregated figures of value added as indicators of firm output rather than use volume figures. It also explains why in one study it was decided to compare and benchmark the efficiency of projects that were of a broadly similar type, that is, projects that comprised three-storey blocks of flats (Ingvaldsen 2005). Regardless of which approach is used to relate outputs

Table 6.1 Examples of papers on construction industry productivity

AUTHORS (DATE)	COUNTRY	METHODOLOGY USED
Kendrick (1956a, 1956b, 1961, 1973)	USA	TFP index
Schultz (1959)	USA	TFP index
Haber and Levinson (1956)	USA	TFP index
Alterman and Jacobs (1961)	USA	TFP index
Dacy (1965)	USA	Production function
Cassimatis (1969)	USA	Partial index approach
Cremeans (1981)	USA	Partial index approach
Stokes (1981)	USA	Production function
Kau and Sirmans (1983)	USA	Production function
Maloney (1983)	USA	Site-level labour productivity
Allen (1985)	USA	Production function
Schriver and Bowlby (1985)	USA	Cost function
Allmon et al. (2000)	USA	Site-level labour productivity
Lowe (1987)	USA	Site-level labour productivity
Orr (1989)	New Zealand	Production function
Chau (1993)	Hong Kong	Cost function
Chau and Walker (1988)	Hong Kong	Partial index
Thomas and Sakarcan (1994)	USA	Site-level labour productivity
Chau and Lai (1994)	Hong Kong	TFP index
Chapple (1994)	New Zealand	Production function
Färe et al. (1996)	New Zealand	DEA
Philpott (1995)	New Zealand	Production function
Wang and Chau (1997, 2001)	Hong Kong	DEA
Diewert and Lawrence (1999)	New Zealand	TFP index
Tan (2000)	Singapore	TFP index
Ive and Gruneberg (2000)	USA	Partial index
Black et al. (2003)	New Zealand	TFP index
Pearce (2003)	USA	Partial index
Goodrum and Haas (2004)	USA	Production function
Chau and Wang (2005)	Hong Kong	DEA
Edvardsen (2005)	Norway (construction firms)	DEA
Ingvaldsen (2005)	Norway (building projects)	DEA
McCabe et al. (2005)	Canada (construction firms)	DEA
Briscoe (2006)	UK	TFP index
El-Mashaleh et al. (2007)	Florida, USA (construction firms)	DEA
Mason and Osborne (2007)	New Zealand	Production function
Choy (2008)	Malaysia	DEA
Xue et al. (2008)	China	DEA
Chau (2009)	Hong Kong	Cost function
Wang, Ye and Yuan (2010)	China	DEA
Chiu and Wang (2011)	Taiwan (construction firms)	DEA
Horta et al. (2012)	Portugal (construction firms)	DEA
Li and Liu (2011)	Australia	DEA
Kapelko and Abbott (2017)	Spain	DEA

and inputs to each other, the construction industry does have some special characteristics that makes the determination of these outputs and inputs difficult.

The construction industry

In the construction industry there are a number of problems that have arisen
in the determination of output measures which are fairly unique to the industry. This has had an impact on the perception of productivity growth, or the
lack of it, in the industry. The growth (or lack of growth) of productivity in the
construction industry in a number of countries has been an issue since the late
1960s. Research in the United States that found a stagnation in productivity
growth includes work by Stokes (1981), Allen (1985) and Schriver and Bowlby
(1985). Tan (2000) found a similar decline in productivity in the Singapore construction industry over the period 1980–1996. Since these works were published,
researchers have tried to understand both why this might have occurred and the
technical difficulties involved in determining levels of productivity in the construction industry. A number of economists have put forward possible reasons for
the stagnant growth of productivity in the industry; these have included such
things as the high labour intensity of the industry, the low economies of scale in
the industry, a lack of competition, regulatory impediments, faulty innovation
and management practice, union restrictions on work practices, poor investment
quality and a low level of skill and training (Davis 2007; Richardson 2014). Perhaps the most common view that has been expressed is that the stagnation has
been caused by the basic character of the industry, in that to a large degree the
construction industry is a labour-intensive one, which means (it has been argued)
that the introduction of new equipment and technology can only increase production levels with a given amount of capital and labour – therefore making
productivity improvements difficult (Allen 1985).

Besides an actual decline in productivity occurring, one possibility is that it is
the measures themselves that are used to determine the levels of productivity in
the industry that might be faulty, and it could be that the industry is experiencing
productivity improvements that are not being detected (this was argued in the
American case when low levels of productivity growth were detected; see Rosefielde and Mills 1979; Schriver and Bowlby 1985). This would certainly seem to
be a possibility given that the industry has, over the long run, experienced a range
of technological changes. In the last few decades, for instance, new tools and
equipment have been introduced that have tended to be labour saving, including the introduction of handheld powered tools (nail drivers, sanders, saws and
drills), and improved lifting and moving machinery (cranes, loaders, earthmov-
ers, graders and forklifts). In addition, new materials and processes have also been
introduced along with a greater use of prefabricated components. The introduction and greater use of new equipment, materials and processes has meant that
it is hard to explain why growth in the productivity of the construction industry
may be sluggish, and researchers should be aware of the problems associated with

the analytical techniques they use before coming to strict conclusions on the industry's productivity.

One of the main difficulties in determining levels of productivity in the construction industry has been determining levels of output. In the construction industry, physical measures are difficult to use because of the varied nature and complexity of buildings, and the traditional financial measure of value added is also problematic as the industry increasingly moves to off-site fabrication. As work moves off-site, work that is done on-site becomes increasingly unskilled assembly and appears less productive. In terms of its heterogeneity, construction involves many different types of projects. The industry itself includes a range of different types of activities including single- and multi-unit housing, apartments, commercial properties, and heavy engineering projects such as roads and bridges. Even within a sub-sector of the industry there can be many variations between different types of seemingly similar projects because of differences in design, architecture, quality and locational factors. This refers to the heterogeneous nature of construction products, both in type and location. Project characteristics, such as the increased size and complexity of projects, resulting communication difficulties and fast-tracking projects where design and construction phases overlap, also affect coordination. Generally, therefore, each construction project is designed and built to serve a special need. Although specific design and construction skills are needed over and over again, the outputs differ in size, configuration, location and complexity. Such uniqueness impacts substantially on construction productivity and the construction process. In addition, the site-based nature of construction and project management is important. Those studies that use aggregated national level data tend not to incorporate these changes except to the degree that they are reflected in the value of the buildings.

In the case of construction, the main indicator of output that is used is that of value added. Other alternatives include such things as the physical area of construction space (gross floor area) or the number of dwellings constructed. What one is struck by when looking at these various measures is how different they can be. Figure 6.1 provides data on indices of output of the building of new residential units (houses and apartments) in New Zealand between 1984 and 2012. Indices for floor space, the number of units and a constant dollar value of value added are provided. Although the three indices do move in the same direction, they indicate quite different levels of output in the industry.

This is by no means a problem that is just associated with the construction industry. It has long been known that there is a problem of measuring output in general, much more so in the services sector than in the goods producing sector. Most measurement problems boil down to the fact that service activities are intangible, are more heterogeneous than goods, and are often dependent on the actions of the consumer as well as the producer. The main problem is the measurement of output volumes, which requires accurate price measurement adjusted for changes in the quality of a service (O'Mahoney and Timmer 2009; Inklaar 2008).

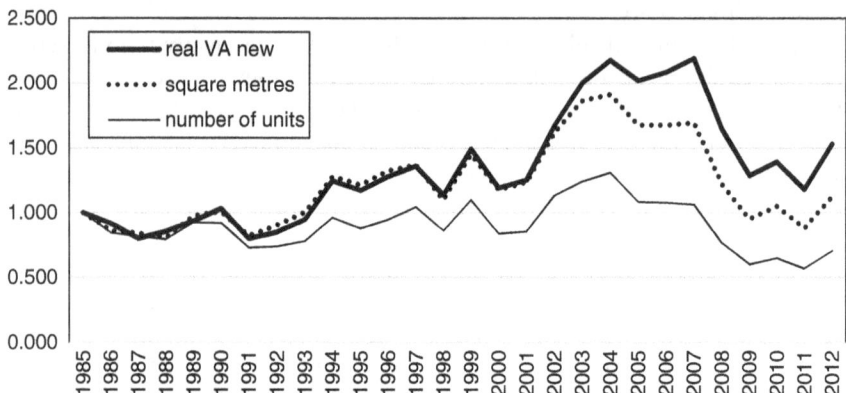

Figure 6.1 Indices of indicators of new residential building in New Zealand (1985 = 1.00), 1985 to 2012

Source: Statistics New Zealand (2011).

The very substantial differences of types of projects contained within the construction industry has been recognized by a number of researchers and been the basis of one view that much of the perceived stagnation in productivity can be attributed to a change in the output mix from high to low productivity building projects (Schriver and Bowlby 1985), that is, changes in the mix of outputs within the industry. Another related explanation proposed is that inappropriate price indices have been used in determining productivity measures, and therefore the perceived stagnation is not as serious as it appears (Dacy 1965; Stokes 1981). The price indices used might be inappropriate because they are based on aggregations of building materials prices that are not weighted according to their use. This is related to the previously mentioned explanation that official data on construction industry activity does not take into account improvements in the quality of the construction outputs produced, and so therefore improvements have not been detected (Rosefielde and Mills 1979).

In a number of studies, output was based on expenditure or the value added of construction from official statistical sources. Few attempts have been made to use other indicators of output such as square metres of floor space, although where these have been used other problems have arisen (Chancellor *et al.* 2015). In these cases, and in cases where expenditure figures are used, they might not capture all of the improvements in the quality of the materials and design of construction projects, or a change in the composition of the industry – residential versus commercial or engineering projects (or indeed other intangibles in the industry such as the improvement of safety standards or times to completion of projects). In some instances, it was probably the case that approaches were chosen because of the ease of collecting the necessary data from official sources. There is nothing especially wrong with this, although it does ignore the issues of quality that were originally raised back in the 1970s and 1980s. It does mean, however, that there is only so much that can be learnt from these studies in terms of identifying the main drivers of growth in productivity in the construction industry.

In addition, the shadow economy (see Chapter 5) is often not considered in productivity research, as its impact on productivity estimation is seen, in most cases, as negligible. Most studies, however, estimate the construction industry as possessing the greatest shadow economy compared to normal production (Schneider 2013; ABS 2013; Chancellor and Abbott 2015). It is possible that some part of the output of the construction industry, especially renovations in the residential building sub-sector, are not being included in aggregated national figures of construction output.

Price indices

Related to the problem of determining measures of output is the use of price indices in deflating values of construction expenditure or value added to constant dollar terms (OECD 2002; Bresnahan and Gordon 1996). In the case of the construction industry it has been argued that construction materials and property price indices do not accurately reflect the input costs and output prices of the industry. One of the reasons, therefore, that the measured rate of construction productivity growth may be low is because the measurement of output as value added (the total value of goods and services produced after deducting the costs in the production process) is adjusted by a deflator for movements in prices that is in some way faulty. The construction deflator may not fully take movements in output prices into account, and therefore real output is underestimated. Output of the construction industry is often estimated by deflating current price figures by input price indices. A number of researchers have criticized the use of input price indices for deflating construction expenditure, for being unrepresentative of the inputs priced and geographical coverage, and for being based on inaccurate weights (Dacy 1965; Stokes 1981).

A number of alternative deflators have been developed. Allen (1985) used a price per square foot index for deflating non-residential building, assuming that this is a good proxy for output. According to Allen's estimates, about half the decline in construction productivity during the 1960s and 1970s was due to the over-deflation of construction output. Cassimatis found that price indices cannot provide adequate deflators for construction: 'the feeling persists that construction productivity is greater than the measurements show . . . largely due to the fact that there are no adequate price indices that can be used as deflators of the gross product' (Cassimatis 1969: 79–80). Pieper (1990) also argued that deflation by input price indices does not produce suitable estimates of output at constant prices and, given the extensive use of input price indices as deflators in estimating the constant price of output for the construction industry, productivity measurement for this industry is problematic, to say the least. Pieper concluded that, for the United States, 'evidence indicates an over deflation of construction of at least 0.5% per year between 1963 and 1982' (Pieper 1990: 239).

This problem has also occurred in the case of countries other than the United States. Chau and Lai (1994), for instance, developed a system for measuring the relative labour productivity of the Hong Kong construction industry. In their approach the trend of construction labour productivity was derived from national accounts data. Relative rates of growth of labour productivity were based on an implicit price deflator for net output of the construction industry obtained through double deflation. The price indices were based on a construction output price index and a material cost index using the methodology developed by Chau and Walker (1988).

In order to overcome some of these problems, attempts have been made to use price indices that apply to very specific areas of the construction industry. Sveikauskas *et al.* (2014), for instance, looked at the building of single- and multiple-family residential construction and the construction of roads in the United States. They devised price indices for these restricted areas only. Interestingly, they found that after these adjustments productivity appeared to have grown in the sector in the 1990s and 2000s.

There are other examples: Lowe (1995) described the use of estimation indices by Statistics Canada, using surveys sent to sub-contractors. Around 100 different items were priced for five building types, and each of the five types had its own index. A recent analysis of British building price indices by Yu and Ive (2008) found that these indices measure the price movement of the traditional building trades but almost completely ignore mechanical and electrical services. Cannon (1994) questioned the accuracy of contractor statistics, and Briscoe (2006) asked: 'How useful and reliable are construction statistics?' These papers identified a range of problems with data collection and analysis, including defining the scope and coverage of the industry, measuring outputs across different types of activity, identifying construction firms, measuring capital formation and capital stock and inconsistent employment statistics. Crawford and Vogl (2006) also drew attention to data limitations for productivity analysis.

Further studies that concentrate on far more specific examples of fairly consistent and comparable building projects need to be undertaken if it is going to be possible to arrive at productivity measurements that researchers can have full confidence in.

Conclusion

Many challenges and opportunities for methodological improvement exist in determining construction output, which at face value appears to be simple. Of course, this is not the case, and many of the difficulties have been presented in the preceding discussion. Historically, the use of basic monetary measures such as gross value added have been popularly applied, perhaps due to ease of data availability and use. Physical measures have also been used, despite their shortcomings. Further work is required to understand construction output, so that more reliable and meaningful estimates of productivity can be produced, which in turn can be made use of in exploring the ways in which construction productivity can be further improved.

References and further reading

ABS (2013) *Information Paper: The Non-Observed Economy and Australia's GDP*, 2012, Cat. No. 5204.0.55.008 (Canberra: Australian Bureau of Statistics), 3 July 2016. www.abs.gov.au/ausstats/abs@.nsf/Latestproducts/5204.0.55.008Main%20Features52012?opendocument&tabname=Summary&prodno=5204.0.55.008&issue=2012&num=&view.

Allen, S. G. (1985) Why construction industry productivity is declining. *The Review of Economics and Statistics*, **67** (4), 661–669.

Allmon, E., Haas, C., Borcherding, J. and Goodrum, P. (2000) US construction labor productivity trends 1970–1998. *Journal of Construction Engineering and Management*, ASCE, **126** (2), 97–104.

Alterman, J. and Jacobs, E. E. (1961) Estimates of real product in the United States by industrial sector, 1947–1955. In: *Output, Input and Productivity Measurement* (Princeton: National Bureau of Economic Research, Princeton University Press).

Black, M., Guy, M. and McLellan, N. (2003) *Productivity in New Zealand 1988 to 2002*. Treasury Working Paper 03/06 (Wellington: The Treasury).

Bresnahan, T. F. and Gordon, R. J. (eds.) (1996) *The Economics of New Goods, Studies in Income and Wealth* (Chicago: University of Chicago Press).

Briscoe, G. (2006) How useful and reliable are construction statistics? *Building Research and Information*, **34** (3), 220–229.

Cannon, J. (1994) Lies and construction statistics. *Construction Management and Economics*, **12** (4), 307–312.

Cassimatis, P. J. (1969) *Economics of the Construction Industry* (New York: National Industrial Conference Board).

Chancellor, W. and Abbott, M. (2015) The Australian construction industry: Is the shadow economy distorting productivity? *Construction Management and Economics*, **33** (4).

Chancellor, W., Abbott, M. and Carson, C. (2015) Factors promoting innovation and efficiency in the construction industry: A comparative study of New Zealand and Australia. *Construction Economics and Building*, **15** (2), 64–81.

Chapple, S. (1994) *Searching for the Heffalump? An Exploration into Sectoral Productivity and Growth in New Zealand* (Wellington: New Zealand Institute of Economic Research Working Paper 99/10).

Chau, K. W. (1993) Estimating industry-level productivity trends in the building industry from building cost and price data. *Construction Management and Economics*, **11** (4), 370–383.

Chau, K. W. (2009) Explaining total factor productivity trend in building construction: Empirical evidence from Hong Kong. *International Journal of Construction Management*, **9** (2), 45–54.

Chau, K. W. and Lai, L. W. C. (1994) A comparison between growth in labour productivity in the construction industry and the economy. *Construction Management and Economics*, **12**, 183–185.

Chau, K. W. and Walker, A. (1988) The measurement of total factor productivity of the Hong Kong construction industry. *Construction Management and Economics*, **6** (3), 209–224.

Chau, K. W. and Wang, Y. S. (2005) An analysis of productivity growth in the construction industry: A non-parametric approach. In: Khosrowshahi, F. (ed.) *21st Annual ARCOM Conference*, 7–9 September 2005, SOAS, University of London, Association of Researchers in Construction Management, Vol. 1, 159–169.

Chiu, C. Y. and Wang, M. W. (2011) An integrated DEA based model to measuring financial performance of construction companies. *WSEAS Transactions on Business Economics*, 1 (8), 1–15.

Choy, C. F. (2008) Productive efficiency of Malaysian construction sector, *Proceedings of 14th Annual Conference of the Pacific Rim Real Estate Conference*, January 20–23, Kuala Lumpur. www.prres.net/papers/chia_productive_efficiency_of malaysian_construction.pdf

Crawford, P. and Vogl, B. (2006) Measuring productivity in the construction industry. *Building Research & Information*, 34 (3), 208–219.

Cremeans, J. E. (1981) Productivity in the construction industry. *Construction Review*, 27 (5), 4–6.

Dacy, D. C. (1965) Productivity and price trends in construction since 1947. *Review of Economics and Statistics*, 47 (4), 406–411.

Davis, N. (2007) *Construction Sector Productivity Scoping Report* (Wellington: Martin Jenkins).

Diewert, E. and Lawrence, D. (1999) *Measuring New Zealand's Productivity* (Wellington: Treasury Working Paper 99/5, Treasury).

Edvardsen, D. F. (2005) Economic efficiency of contractors. In: Kaehkoenen, K. and Porkka, J. (eds.) *Global Perspectives on Management and Economics in the AEC Sector, Volume II*. 11th Joint CIB International Symposium Combining Forces – Advancing Facilities Management and Construction through Innovation, 13 June 2005–16 June 2005, Helsinki.

El-Mashaleh, M. S., Minchin, R. E. and O'Brien, W. J. (2007) Management of construction firm performance using benchmarking. *Journal of Management in Engineering*, 23, 10–17.

Färe, R., Grosskopf, S. and Margaritis, D. (1996) Productivity growth. In: Silverstone, B., Bollard, A. and Lattimore, R. (eds.) *A Study of Economic Reform: The Case of New Zealand* (New York: Elsevier Science).

Ganesan, S. (1984) Construction productivity. *Habitat International*, 8, 3–4.

Goodrum, P. M. and Haas, C. T. (2004) Long-term impact of equipment technology on labor productivity in the U.S. construction industry at the activity level. *Journal of Construction Engineering and Management*, 130 (1), 124–133.

Haber, W. and Levinson, H. (1956) *Labor Relations and Productivity in the Building Trades* (Ann Arbor: University of Michigan Press).

Horta, I. M., Camanho, A. S. and Moreira da Costa, J. (2012) Performance assessment of construction companies: A study of factors promoting financial soundness and innovation in the industry. *International Journal of Production Economics*, 137 (1), 84–93.

Ingvaldsen, T. (2005) Scientific benchmarking of building projects – model and preliminary result. In: Kazi, A. S. (ed.) *Systematic Innovation in the Management of Construction Projects and Processes, Volume III*. 11th Joint CIB International Symposium: Combining Forces – Advancing Facilities Management and Construction through Innovation, 13–16 June, Helsinki.

Inklaar, R. (2008) The sensitivity of capital services measurement: Measure all assets and the cost of capital, Working Paper GD-103, Groningen Growth and Development Centre, University of Groningen.

Ive, G. and Gruneberg, S. (2000) *The Economics of the Modern Construction Sector* (Basingstoke: Palgrave Macmillan).

Ive, G., Gruneberg, S., Meikle, J. and Crosthwaite, D. (2004) *Measuring the Competitiveness of the UK Construction Industry* (London: Dept. of Trade and Industry).

Kapelko, M. and Abbott, M. (2017) Productivity growth and business cycles: The case of the Spanish construction industry. *Journal of Construction Engineering and Management*, **143** (5).

Kau, J. B. and Sirmans, C. F. (1983) Technological change and economic growth in housing. *Journal of Urban Economics*, **13**, 283–295.

Kendrick, J. W. (1956a) *Productivity Trends: Capital and Labor*, Occasional Paper 53 (Washington, DC: National Bureau of Economic Research).

Kendrick, J. W. (1956b) *The Meaning and Measurement of National Productivity* (Washington, DC: PhD thesis, George Washington University).

Kendrick, J. W. (1961) *Productivity Trends in the United States* (New York: National Bureau of Economic Research, Princeton University Press).

Kendrick, J. W. (1973) *Postwar Productivity Trends in the United States 1948–1966* (New York: National Bureau of Economic Research).

Kendrick, K. W. and Jones, C. E. (1951) Gross National Farm product in constant dollars, 1910–1950. *Survey of Current Business*, **31**, September, 12–19.

Li, Y. and Liu, C. (2011) Construction capital productivity measurement using data envelopment analysis. *International Journal of Construction Management*, **11** (1), 49–61.

Lowe, J. G. (1987) The measurement of productivity in the construction industry. *Construction Management and Economics*, **5**, 115–121.

Lowe, P. (1995) Labour-productivity growth and relative wages: 1978–1994. In: Andersen, P., Dwyer, J. and Gruen, D. (eds.) *Productivity and Growth: Proceedings of a Conference* (Sydney: Reserve Bank of Australia).

Maloney, W. (1983) Productivity improvement: The influence of labor. *Journal of Construction Engineering Management*, **109** (3), 321–334.

Mason, G. and Osborne, M. (2007) *Productivity, Capital-Intensity and Labour Quality at Sector Level in New Zealand and the UK* (Wellington: New Zealand Treasury Working Paper 07/01).

McCabe, B., Tran, V. and Ramani, J. (2005) Construction prequalification using data envelopment analysis. *Canadian Journal of Civil Engineering*, **32**, 183–193.

OECD (2001) *Measuring Productivity: OECD Manual Measurement of Aggregate and Industry Level Productivity Growth Organisation for Economic Co-operation and Development* (Paris: OECD Publications Service).

OECD (2002) *Measuring the Non-Observed Economy – A Handbook Organisation for Economic Co-operation and Development* (Paris: OECD Publications Service).

O'Mahoney, M. and Timmer, M. P. (2009) Output, input and productivity measures at the industry level: The EU KLEMS database. *Economic Journal*, **119**, F374–F403.

Orr, A. (1989) *Productivity Trends in New Zealand: A Sectoral and Cyclical Analysis 1961–1987* (Wellington: NZIER).

Pearce, D. (2003) *The Social and Economic Value of Construction* (London: Davis Langdon Consultancy).

Philpott, B. (1995) *New Zealand's Aggregate and Sectoral Productivity Growth 1960–1995* (Wellington: Research Project on Economic Planning, Paper 274, Victoria University of Wellington).

Pieper, P. E. (1990) The measurement of construction prices: Retrospect and prospect. In: Berndt, E. R. and Triplett, J. E. (eds.) *Fifty Years of Economic Measurement* (Chicago: National Bureau of Economic Research, University of Chicago Press), 239–272.

Richardson, D. (2014) *Productivity in the Construction Industry* (Canberra: Technical Brief No 33: The Australia Institute).

Rosefielde, S. and Mills, D. Q. (1979) Is construction technologically stagnant? In: Lange, J. E. and Mills, D. Q. (eds.) *The Construction Industry* (Lexington, MA: D.C. Heath and Company).

Schneider, F. (2013) *The Shadow Economy in Europe, 2013* (Linz: Johannes Kepler Universität).

Schriver, W. R. and Bowlby, R. L. (1985) Changes in productivity and composition of output in building construction, 1972–1982. *Review of Economics and Statistics*, **67** (2), 318–322.

Schultze, C. L. (1959) *Prices, Costs and Output for the Post War Decade: 1947–1957* (New York: New Committee for Economic Development).

Stokes, H. K. (1981) An examination of the productivity decline in the construction industry. *Review of Economics and Statistics*, **63** (4), 495–502.

Sveikauskas, L., Rowe, S., Mildenberger, J., Price, J. and Young, A. (2014) *Productivity Growth in Construction* (Washington, DC: BLS Working Papers, US Bureau of Labor Statistics).

Statistics New Zealand (2011) *Industry Productivity Statistics, 1978–2009* (Wellington: Statistics New Zealand).

Tan, W. (2000) Total factor productivity in Singapore construction. *Engineering, Construction and Architectural Management*, **7** (2), 154–158.

Thomas, H. and Sakarcan, A. (1994) Forecasting labor productivity using factor model. *Journal of Construction Engineering Management*, **120** (1), 228–239.

United Kingdom, Department of Business Enterprise and Regulatory Reform (2008) *Industry Performance Report 2008* (London: Department of Business Enterprise and Regulatory Reform).

Wang, Y. S. (1998) *An Analysis of the Technical Efficiency in Hong Kong's Construction Industry* (Hong Kong: Unpublished PhD thesis, University of Hong Kong).

Wang, Y. S. and Chau, K. W. (1997) *An Evaluation of the Technical Efficiency of Construction Industry in Hong Kong Using the DEA Approach* (Cambridge: Proceedings of ARCOM97 Conference), 690–701.

Wang, Y. S. and Chau, K. W. (2001) An assessment of the technical efficiency of construction firms in Hong Kong. *International Journal of Construction Management*, **1** (1), 27–29.

Wang, H., Ye, G. and Yuan, H. (2010) *An AHP/DEA Methodology for Assessing the Productive Efficiency in the Construction Industry* (Wuhan, China: Management and Service Science 2010, International Conference), 24–26 August.

Xue, X., Shen, Q., Wang, Y. and Lu, J. (2008) Measuring the productivity of the construction industry in China by using DEA-based Malmquist productivity indices. *Journal of Construction Engineering Management*, **134** (1), 64–71.

Yu, M. K. W. and Ive, G. (2008) The compilation methods of building price indices in Britain: A critical review. *Construction Management and Economics*, **26** (7), 693–705.

Editorial comment

Productivity has many definitions, with the simplest being the familiar 'output divided by input', or 'the ratio of output (product) and the inputs required to produce that output'. These deceptively simple definitions mask the myriad complexities that are associated with trying to measure productivity.

One fundamental problem is the identification of exactly what the inputs and outputs of an industry or even an individual firm are. Very few, if any, producers deliver just a single product; car makers, for example, produce a range of models. In these days of mass customisation, even cars of the same model can be ordered with bespoke add-ons and accessories, with the result that individual vehicles are subtly different.

Defining and measuring output in construction is particularly difficult as construction is notorious for the diversity of its products, which range from single dwellings to high-rise apartments to infrastructure, and it includes both newbuild and refurbishment plus other work on existing buildings/structures and more. Construction inputs comprise the usual labour and materials, but capital inputs in various forms such as machinery and equipment and training of employees must also be considered. It is the issue of productivity and, particularly, productivity improvement associated with capital inputs, which is the subject of this chapter.

Improving productivity is not simply a matter of making people work harder – even if it were, there must be limits to how much a single worker can do in a given period of time. If, however, a worker gains more skills through training and/or is provided with a piece of equipment that increases their output then productivity can be boosted, but this requires investment (i.e. inputs of capital). Once we look at levels above that of the individual worker in construction, there are many more possibilities that can be considered and investigated – research and development (R&D), for example – that can lead to improved materials and better work methods for the incorporation of such materials into structures. This may include off-site fabrication as well as more efficient on-site processes. Measuring the impact of investment in R&D is just one of many considerations in assessing capital productivity in the construction industry. In this chapter, the author addresses a wide range of factors that affect capital productivity and uses the Australian construction industry as a case study to illustrate aspects of the capital productivity puzzle.

7 Measuring capital productivity in construction

Michael Regan

Introduction

Productivity is the ratio of output to input in the production process and is a measure of the productive efficiency of the economy. Labour productivity measures output generated for each unit of labour input and, like capital productivity, is considered a partial productivity measure because of its reliance on a single input. Labour productivity is calculated for the economy as real gross domestic product (GDP) per hour worked.[1] Capital productivity estimates are indicators of real GDP per unit of capital inputs or services used in production. Multifactor productivity (MFP) measures the amount of real output expressed in real value added from inputs of capital and labour.

The most comprehensive measure of an economy's productivity is MFP, which is the efficiency with which producers generate additional output from inputs of capital and labour (Productivity Commission 2015: 2). Growth in labour productivity captures improved labour efficiencies as well as the value added from growth in capital productivity through such mechanisms as research and development, technical progress and technology embodied in new plant and equipment. These environmental factors or externalities provide incremental output without the use of additional labour inputs. Investment in technical progress is central to both labour and capital productivity and directly affects MFP. Other externalities on the input side are the pricing of inputs, currency exchange rates (Productivity Commission 2016), public capital expenditures (Pereira and Roca-Sagales 1999; Abiad *et al.* 2015) and the utilisation of capital (Barnes 2011).

On the output side, externalities in the case of commodity exports include exchange rate or commodity price volatility, shifts in demand or a downturn in the business cycle. If investment continues in order to complete projects commenced several years earlier and the value of output diminishes, productivity will decline. Other matters that may affect productivity on the output side include externalities such as mismatch of business and investment cycles (KPMG 2016) and change in the terms of trade (Productivity Commission 2014a, 2016).

MFP is calculated within a growth accounting framework based on Solow's seminal work on neoclassical growth theory (Solow 1956). The framework uses an aggregate production function with one output and two inputs, labour and capital. In this framework, MFP measures changes in technology captured in capital service and intermediate inputs. The difference between input and output values is the producer surplus shared with consumers of the good or service through the market mechanism. Data is represented as an index for the market sector of the economy and is monitored for short, medium and long-term trends, and to identify differences between industries (ABS 2007: 99). In the neoclassical growth framework, capital accumulation drives growth in the short term but experiences diminishing returns over time, which suggests capital productivity in the long term is due to exogenous technical progress – exogenous in the sense that this factor lies outside the growth model (Stiroh 2001).

The growth accounting framework is designed to show how much real growth in output is derived from increased use of capital and labour inputs and how much originates from productivity improvements (Productivity Commission 2013: 4). Growth accounts and productivity estimates are prepared for productivity growth cycles by calculating the average annual growth rate for MFP between cyclical peaks. International comparisons using this approach are problematic because of differences between countries in economic and productivity growth cycles, industry structure, exposure to exchange rate volatility and international competitiveness.

Economists and policymakers argue that it is necessary to take a long-term view of MFP, which is subject to vagaries in the growth of the business cycle and changes in capacity utilisation (Australian Parliament 2010: 25). However, this is only possible at the industry level because of limitations which do not permit long-term productivity measurement at the firm level (Barnes 2011).

A further measure of productivity is capital deepening, which refers to a change in the amount of capital available per unit of labour (Productivity Commission 2016). It is a measure of the difference between growth in labour productivity and MFP. An improvement in the ratio of capital to labour is generally correlated with improved labour productivity (Parham 2013).

Productivity is important. It contributes to improved living standards and international competitiveness, as well as driving increases in real incomes and improvements at the enterprise level (Productivity Commission 2014b; Eslake and Walsh 2011). Long-run productivity performance is influenced by environmental factors such as economic and social institutional frameworks, a robust market economy, trade and investment openness, efficient industry regulation, and government policies that favour competition in the market sector, openness to trade and investment, and microeconomic reforms that lower transaction costs at the firm level. These are formidable challenges if countries in the Organisation for Economic Co-operation and Development (OECD) are to improve growth in real income per capita in coming decades. While governments can introduce policy and incentive frameworks to create a favourable environment for productivity change, ultimately performance depends on the actions of individual firms adopting best practice management standards that will achieve greater efficiencies at the enterprise level (Dowdy and Van Reenen 2014; Solow 2014).

This chapter provides a review of empirical evidence that explains the drivers of capital productivity in OECD economies and the role that capital productivity plays in both labour and multifactor productivity performance with reference to the construction industry. A limitation of the review is data incompatibility between national measures of national capital stock, and sampling and pricing techniques used to calculate gross value added indicators that measure capital productivity output. A case study approach is adopted with a review of productivity performance, with reference to OECD benchmarks, where this provides comparative insights.

Measuring productivity in OECD countries

Guidelines for the estimation of productivity data were adopted by the OECD in 2001. Australia has followed OECD guidelines, although periodically experimental estimates are prepared to provide deeper analysis of the drivers of growth at the industry level (ABS 2007, 2015e). A number of countries recently adopted new methods for estimating productivity, most notably the KLEMS[2] system now used in the United States, Canada, Japan, Korea, the European Union, India and China. While retaining the growth accounting framework, the principal change in the KLEMS system is the extension of the value added estimation by the inclusion of intermediate inputs, which is the value of goods and services consumed as inputs in the production process (ABS 2015e; Jorgenson *et al.* 2007).

Measuring MFP is problematic because of the absence of industry-level data for the preparation of long-run estimates and the large number of factors that create deviations from industry trend over time (Barnes 2011). For example, productivity is measured for Australia's market sector (excluding communications, finance and insurance and cultural services), which in the 10 years to 2015 accounted for 56% of the nation's GDP (Productivity Commission 2016). Two measures are used for labour inputs, hours worked and quality adjusted hours worked, which takes into account changes in labour composition brought about by education, training and work experience (ABS 2015c). The estimates also measure paid and unpaid overtime, weekly hours, employed and self-employed persons, annual leave and the proportion of part-time employees. Labour productivity is based on hours worked by industry and output using gross value added (GVA) from national GDP data. The labour productivity index is estimated by dividing the index of the volume of GVA by an index of labour input (ABS 2007).

Capital inputs or services are more difficult to measure, with data limitations at the industry and firm level. Estimates are based on selected asset groups for which average values for investment, retirements and average age are calculated using the perpetual inventory method. This approach encounters a heterogeneity problem, which occurs when assets of different ages, operating in different industries, with various levels of utilisation and different rates of wear and tear are treated as though they were one. The capital services index is calculated by weighting assets with notional rental prices. International comparisons for capital productivity are difficult to calculate for OECD countries because of differences in methods used to deflate information and communication technology (ICT) investment, issues with the pricing of constant quality price changes, and individual country approaches to calculating net capital stock, asset depreciation and productive service lives (OECD 2016a: 50).

There are several major flaws in the present method of calculating capital productivity. The neoclassical growth model relies on a number of key assumptions which include reliance on a static production function with a Cobb-Douglas specification, and that all firms are profit-maximising and operate in perfect competition with constant returns to scale. Neither assumption stands close scrutiny.

First, the limitations of the production function model have been widely acknowledged in growth economics for over 30 years, particularly its susceptibility to simultaneity bias (Pereira and Alvarez 2013). More recent growth research has made wider use of endogenous growth theory and dynamic multivariate vector autoregressive models, which take into account other variables in the economy and have been shown to address causation concerns with more accurate outcomes.

Second, the assumption of constant returns to scale is implausible. Economies of scale are cost savings that are the result of growth in the size of an enterprise, the scale of operations and the extent to which firms have engaged in vertical and horizontal integration. In practice, marginal costs diminish as fixed costs spread over more units of output and returns improve. Operational efficiencies also improve with scale, and lower variable costs further contribute to improved returns. Foster (2015) posits that improved MFP performance is being achieved by increasing returns to scale made possible by investment in capital goods, improved workforce skills, embodied technology and rapid growth in network connections with subsequent advantages to scale. Others have observed a similar phenomenon in the United States, with economies of scale assuming greater importance to growth than MFP (Diewert and Fox 2008).

International productivity performance

In the past 10 years, many OECD countries experienced negative growth. There were several reasons for this, including a decline in public and private investment, the need to correct fiscal deficits and public debt in the aftermath of the global financial crisis, the international boom in the resources sector and cyclical variation in factor utilisation as a consequence of change in industry structure (OECD 2016a). While significant differences exist in the productivity performance of OECD countries, cross-country GDP per capita performance can be attributed to differences in labour productivity growth, high unemployment and a decline in labour utilisation. MFP growth and capital deepening slowed in most OECD countries following the crisis with negative impacts on labour productivity.

There has been a steady decline in labour productivity growth and capital intensity in the G7 countries since the early 1970s. While capital productivity has remained negative since the mid-1990s, investment in the knowledge-based industries is high and increasing at a faster rate than investment in tangible assets, particularly in information and communication services, technology adoption, international network connections and supply chains. Information and communications technology accounted for an average 17.5% of labour productivity growth in 2014 (OECD 2016a: 17, 22).

The construction industry has recorded mixed results in the OECD, with wide performance differences between countries (OECD 2016a: 61). In the period 2001–2014, the construction industry made a negligible contribution to growth in market sector GVA (OECD 2016a: 63).

Productivity performance in Australia

Australia experienced three distinct economic growth phases over the past 25 years: 1974–1994, or the baseline period; the growth surge of 1995–2004; and the slump of 2005–2012. In both the baseline and slump cycles, output only just managed to keep pace with input, delivering a small MFP gain of 0.7% and then a downturn of −0.4% in the slump cycle (Parham 2013; ABS 2015b).[3] Capital productivity remained high during the slump cycle, mainly because of the lag in investment following completion of several major gas and mining projects. In the surge cycle, output exceeded combined inputs for an MFP dividend of 1.8% per annum. The slump cycle was also a time of strong growth in capital investment led by the mining industry, with capital intensity per hour worked growing at 3.8% annually between 2005 and 2012, while labour productivity growth declined to its lowest level in 25 years (Taylor *et al.* 2012).

Growth in output is also central to productivity improvement, and during the growth cycle between 1989 and 2004, real output growth of 3.6% exceeded growth in inputs of 2.7% to generate an MFP gain of 0.7% (ABS 2012). Over the past 20 years, Australia experienced growth in both labour (hours worked) and capital inputs which is reflected in the change in the total inputs index (58.7 to 101.9) over this period. Capital deepening is expressed as an index, and between 2005 and 2015 the index value for capital deepening increased from 46.8 to 102.5 (ABS 2015c).

Several explanations are advanced for the productivity slump cycle of 2005–2012. First, Australia experienced favourable terms of trade in both the surge and the slump cycles, largely as a result of the mining boom and high commodity prices. These factors contributed to the maintenance of real income and an average GDP growth rate of 3.2% per annum between 1974 and 2012 (ABS 2013). Second, input growth increased significantly during the slump period led by the mining, manufacturing and construction industries. The mining industry accounted for 8.6% of GDP in 2015, generated AUD 356 billion in revenue and employed 1,038,000 persons (ABS 2015a). The strong rise in commodity prices provided an incentive for miners to increase investment and employment through expansion of existing operations and the recommissioning of marginal projects now justified by improved economics.

A further and important factor in the productivity growth phase was the terms of trade, which declined from its index peak of 118 in 2011–2012 to 90 in 2015,

Table 7.1 Productivity by industry, Australia, 2014–2015 (annual percentage change)

Industry Group	Productivity by Industry 2014–2015				Labour
	MFP	GVA	Labour	Capital	Productivity
			Input	Input	
Agriculture	1.2	1.5	−1.1	1.0	2.6
Mining	5.5	7.6	−12.0	7.4	22.4
Manufacturing	−0.5	−1.2	−0.2	−1.5	−1.0
Utilities	2.5	1.4	−5.9	1.6	7.8
Construction	−2.3	−0.7	0.1	5.6	−0.8
Wholesale trade	0.9	2.5	1.0	2.6	1.5
Retail trade	−0.6	2.6	2.5	5.1	0.1
Accommodation	2.0	7.0	6.0	1.1	0.9
Transport	−3.9	−0.9	3.8	2.3	−4.5
Financial services	4.0	4.6	−1.4	1.8	6.1
Professional services	−4.3	−4.0	−0.5	3.5	−3.5
Cultural services	−4.3	3.0	9.9	3.1	−6.3
Market sector	0.3	2.2	0.8	3.3	1.3

Source: ABS 5260.0.55.002.

leading to a decline in the value of outputs and negative implications for multi-factor growth (Productivity Commission 2016). Recent data indicates that the Australian construction industry in 2014–2015 recorded high capital inputs and no change in labour inputs and made a negative contribution to market sector GVA and MFP (see Table 7.1). The low productivity of the construction industry is due in part to the downturn that followed the mining boom of 2003–2013 (Edwards 2014) and the dispersed nature of the industry in Australia and most OECD countries.

The Australian construction industry

The construction industry is a major component of the national economy, accounting for around 10% of GDP and employing around a tenth of the workforce. In comparison with other industries, construction possesses a number of unique characteristics which partly explain the role of capital productivity in the industry and the challenges for improving future MFP performance in one of the globe's largest industries.

Industry fragmentation

In 2014–2015, the Australian construction industry revenue of AUD 388 billion was derived from three activities: construction services (43%), building construction (35%) and heavy engineering and civil (23%) (ABS 2016c). The industry accounts for an average 13.4% of market sector GVA (ABS 2016a) and employed 1,088,700 persons in May 2016 (ABS 2016a, 2016b). In most OECD countries, including Australia, the industry is composed of a large number of actors with fewer than 5% of registered construction firms employing 500 workers or more and more than 90% of firms employing five persons or fewer. The industry is cyclical and dependent on the business cycle, demand in the housing sector, funding for infrastructure by all levels of government and domestic economic settings.

The Australian construction industry is subject to a patchwork of regulatory arrangements requiring compliance with both international and national standards, state government builder registration and compliance frameworks, and national employment laws (Furneaux *et al.* 2007).

Capital intensity

The Australian construction industry is the least capital intensive in the market sector (ABS 2015b; Lane and Rosewall 2015). In 2015, the value of capital stock per worker was less than AUD 50,000 and labour productivity to GVA per worker was around AUD 132,000 – mid-range for the market sector. While construction industry capital productivity outperformed that of the market sector over the past decade, low capital stock combined with low labour productivity growth resulted in a low contribution to value added per worker (Richardson 2014).

The low capital stock investment by the industry is a result of a number of factors including the practice of accounting for major plant expenditures as an expense of the project, the wide use of short-term equipment hire and subcontractor services, and the high depreciation rate of net capital stock. In the five years to 2014, consumption of capital accounted for 67% of new investment in buildings, plant and equipment (ABS 2015c).

Forms of contracting

In contrast to most other market sector industries, the Australian construction industry makes wide use of contractors, sub-contractors and self-employed tradespeople under short-term project-specific contracts. This is true of the industry in most OECD countries. Under traditional procurement arrangements, the majority of client and contractor relationships are adversarial, with low levels of formal collaboration and limited use of incentives (Regan *et al.* 2015). The Australian industry is experiencing significant change at the present time with wider use of relationship contracting, hybrid procurement methods, partnering principles and the "gateway" project implementation process also adopted in many OECD countries. These reforms have positive implications for industry productivity in the future, although implementation of these reforms is slow and traditional adversarial arrangements still account for most contracts in the industry at the time of writing.

Business planning

The Australian construction industry, like many service industries in the market sector, is subject to the business cycle and the complexity that this introduces for short- and medium-term planning. The unprecedented long run of sustained growth in the Australian economy has underpinned a steady flow of work for the industry assisted by national and state infrastructure programs and the boom in the mining sector in the five years following the global financial crisis. Nevertheless, construction is subject to uncertain work flows and this is a systemic problem for contractors at the firm level, and it explains the industry's short-term approach to investment in human and physical capital and business planning. Around 34% of engineering and construction activity is commissioned by government agencies and is subject to fiscal policy constraints and short-term budgetary and electoral cycles (ABS 2016c). The cyclical nature of construction activity is a major constraint on output (Lane and Rosewall 2015). With a low capital:labour ratio, the industry is lightly capitalised with limited vertical integration and diversification, a parallel situation to that experienced in the UK in the late 1990s (Egan 1998). Historically, most growth in the industry was organic or by specialisation, although this changed in the early 2000s with an international shift to economies of scale and consolidation (Runeson and de Valence 2009). In Australia, the industry has experienced a decade of merger and acquisition activity, with the Oceania region accounting for 38% of global transactions by value in 2015 (PWC 2016). Such merger and acquisition activity has been shown to have no significant impact on firm performance and limits opportunity for medium-term strategic and business planning (Choi and Russell 2004).

Construction performance

The construction sector in most developed countries has a modest performance record, with cost overruns and late delivery of projects the norm rather than the exception (National Audit Office 2003, 2005; Queensland Audit Office 2014). In Australia, a study of 54 projects found that between the original approval and final cost, 35.3% of traditional contracts experienced cost overruns and 25.6% were delivered late. Contracts expected to cost AUD 4.53 billion at contract execution experienced overruns of AUD 672 million or 15% of project value (Infrastructure Partnerships Australia 2007).

Construction performance is also affected by the "winner's curse" in competitive tenders with contractors reducing margins to win contracts with the expectation of being able to recover margin through client-initiated changes in scope or specification during the term of the contract. Traditional procurement practices result in misalignment of incentives between the parties, which suggests that competitive tenders may discourage collaboration between contractors, clients and other participants in the supply chain (Regan *et al.* 2015).

Measuring construction performance has been a long-standing issue for the industry, with the Egan Report (1998) drawing attention to the lack of quantitative information held by firms that could be used to evaluate the success or otherwise of construction projects. This information is necessary to show whether completed contracts have achieved improvements by reducing defects, improving user and client satisfaction, reducing cost and improving efficiency (National Audit Office 2001: 49). Egan found that the UK construction industry had low profitability, and that firms were lightly capitalised and spent little on research and development and training (1998: 4). Traditionally, construction performance is measured using completion time and cost outcomes benchmarked to *ex ante* project objectives. From a client perspective, the objectives are different and include value for money, timely delivery, build quality and fitness for purpose (Latham 1994: 12). Recent evidence suggests that construction performance has improved in OECD countries, although cost and time overruns remain a systemic problem for the industry (Infrastructure Partnerships Australia 2008).

Productivity performance

The construction industry is a major industry in developed and developing countries and has a long record of poor productivity performance identified in reports in the 1990s in the UK (Latham 1994; Egan 1998; National Audit Office 2001), Australia (Productivity Commission 1991; Access Economics 1999; Productivity Commission 2014b; Deloitte Access Economics 2014), New Zealand (NZIER 2013) and Canada (Dozzi and AbouRizk 1993). In Australia, the industry contributed 1.2% per annum in MFP growth between 1990 and 2015, above the market sector average of 0.9% per annum (Productivity Commission 2016). While the industry contributed 9.8% share of market sector GVA and capital inputs were highest in the market sector in 2015, both labour and MFP productivity declined suggesting diminishing returns to capital and lower growth in output value (Productivity Commission 2015). Further evidence of the weakness of the present MFP model is the financial performance of the construction industry in the 10 years to 2013, during which the labour share of GVA declined from 72% to 70% and the profit share increased from 21% to 30%. In 2013, the construction industry return on investment was 107% compared with a market sector average of 28.6% (Richardson 2014). The data suggests that payments to labour were not a significant factor in industry economics and the main beneficiaries of high industry profitability over this period were firm shareholders.

Capital productivity: the drivers

Improving capital productivity is an ongoing challenge across most industries in the market sector in OECD countries. Early work in neoclassical growth theory was based on the principle that permanent and continuous improvements in per capita income were achieved by raising savings, population growth and investment and the introduction of new technology (Harrod 1939; Domar 1946). Subsequent

work by Solow on economic growth and its causes substituted a production relationship in the Harrod-Domar model whereby capital and labour were substitutes.
This change recognised the role of diminishing returns – that is, the productivity
of capital is variable and when the capital:labour ratio rises, diminishing returns
will reduce the productivity of capital (Solow 1956). The research suggests that
capital formation, a change in the capital:labour ratio and population growth are
short-term drivers of improved productivity, and without continuing improvements in technology, productivity will diminish (Regan 2007).[4]

Research by Parham (2013) argues that Australia's recent productivity performance was influenced by input accumulation, MFP growth and favourable shifts
in the terms of trade. The more sustainable of these factors is input accumulation
or growth in the use of capital and labour. However, viewed in a wider context,
the drivers of growth in real income in present market conditions are the available
technology and the efficiency of production within firms and industries (D'Arcy
and Gustafsson 2012). Secondary influences on productivity performance are
volatility and cyclical effects in the economy, changes in industry composition,
changes in the structure of the economy over time, industry regulation, adjustment pressures and measurement errors (Parham 2013). Many of these factors
can be better identified in the context of the wider economy while others may be
determined at the firm level.

The economy level

Investment

Improvement in productivity requires investment in capital inputs and progressive improvement in the capital:labour ratio or capital deepening. Investment is
also required to meet the cost of research and development, employee training
and professional development, and to accelerate procurement programs to gain
the benefits of embodied technology. Most of these activities also require various levels of support from government. Increasing capital inputs contributes to
improved capital productivity when output grows at a faster rate. The relationship
is strongest between productivity and investment in plant and machinery (De
Long and Summers 1991). Investment is recorded as gross fixed capital formation
in national accounts and is a leading indicator of future output capacity, growth
and employment in the economy. The literature widely acknowledges the important contribution that investment plays in productivity and economic development in OECD countries (OECD 2003: 17; D'Arcy and Gustafsson 2012: 29).

In the market sector, investment is used to acquire plant, buildings, civil
works and equipment, and technology, and to improve the skills and know-how
of human capital, particularly in the service industries. The decomposition of
GDP per capita growth shows that both labour productivity and employment
levels explain differences between countries and firms investing in skills training, research and development, innovation and the commercialisation of new
technologies. The highest returns are in the knowledge-intensive industries of

mining, manufacturing, wholesale and retail trades and the media. For investment to improve MFP, there needs to be spillover effects or disembodied technical change that generates additional output for given capital inputs (D'Arcy and
Gustafsson 2012: 29).

Investment flows in Australia in the market sector increased by an average
8.5% annually in nominal terms in the 10 years to 2015, with growth strongest
in mining, the retail industry and construction (Productivity Commission 2016).
Adjusted for mining, nominal growth in the economy was 2%, which was less
than the rate of inflation over this period. Given the high rate of capital retirements in national trading stock data, investment in the market sector declined in
real terms over this period, although there were significant differences between
industries.

The dominant role of mining investment in Australia can be seen in its share
of aggregate investment, accounting for 16.2% of market sector investment in
2005, 34.7% in 2010 and 45.5% in 2015 (ABS 2014a). While mining investment peaked in 2012, residual completion work continued until 2015, during
which time non-mining investment accounted for 5.5% of GDP, which is low
by historical standards but close to trend over the preceding 25 years (Kent
2014). The construction industry experienced strong investment growth in the
10 years to 2015 on an annualised basis and improved capital productivity until
2006 before declining in line with average market sector performance between
2007 and 2015 (Parham 2013; Productivity Commission 2016). However,
while construction investment outperformed the market sector, the industry
has low capital intensity and, along with cultural services, has the lowest net
capital stock values in the market sector.

Investment is a cyclical indicator that experiences high levels of volatility, and
flows may be influenced by economic cycles and other macroeconomic conditions,
business confidence and capacity utilisation across the wider economy. Profitability
and expected future returns are primary determinants of the level of investment.

Capacity utilisation

Capacity utilisation refers to the extent to which existing assets, including
buildings, civil works, plant and equipment, are fully utilised in the economy
and, like investment, capacity utilisation is a volatile indicator determined by
underlying demand and supply pressures in the wider economy. In a period of
stable growth in the economy and high utilisation of productive capacity, firms
in capital-intensive, goods-related industries will hire more labour and invest
in capital stock, which is the inventory of a firm's durable assets. For the services industries, the response is to hire more labour, suggesting that movement
in indicators of utilisation are more likely to provide information about the
labour market than about intentions to invest in the near future. The Australian Industry Group survey in June 2015 (NAB 2016) shows capacity utilisation
in Australia at 71% for the manufacturing industry and 73% for the weighted

average of the services, manufacturing and construction sectors (Lane and Rose-wall 2015). The broad-based quarterly National Australia Bank survey includes all non-farm industries, and in December 2015 showed utilisation across all sectors at 81% (NAB 2016).

Capacity utilisation at the firm level in the capital-intensive industries is based on the maximum output that can be extracted from existing capital stock and labour resources. The service-based industries generally use labour to assess both capacity and utilisation. Some other industries use minimum average cost of production, hours of operation of capital stock, output, availability of skilled labour, and revenue metrics (Lane and Rosewall 2015). However utilisation is measured, firms in the market sector will only invest when confident of achieving hurdle rates of return in the medium term. Capacity constraints are generally regarded as an impediment to long-term growth prospects in industry and the wider economy.

Capital stock

Capital stock refers to aggregate national investment in buildings, intellectual property, plant and equipment.[5] It is calculated annually based on the perpetual inventory method using year-end capital stock values, gross fixed capital formation and capital retirements over a 12-month period. An estimate is also made of the average age of capital stock for each industry group as well as form of ownership (ABS 2016a).

Capital spending in Australia for the 10 years to 2015 showed an average annual increase of 8.7% at current prices with investment strongest in the mining, construction and retail industries. The levels of investment over this period contributed to a decline in the average age of the nation's capital stock from 17.8 to 17.1 years. However, there are significant differences at the industry level, with higher growth in the construction industry of 12.3% per annum compared to cultural services (8.2%), utilities (5.6%), and mining (4.2%). For the five years to 2015 the picture is very different, with mining experiencing 15.2% annual growth, the retail sector 10.3%, construction 6.2% and utilities −6% (ABS 2015c). The large investment in new plant and equipment in the mining industry reduced the average age of capital stock from 9.1 years to 6.7 years in the period 2010–2015.

Australia's market sector capital stock increased at an average 10.5% per annum between 2005 and 2015, although the consumption of fixed capital accounted for an average 66.6% of all new investment. This data provides an alternative perspective on capital productivity and suggests that estimates may be less accurate than present methodology provides. The ratio of capital retirements to investment is highest in manufacturing (124%), financial services (88.7%) and the accommodation sector (84.9%). The ratio for the construction industry was 65%, similar to the average for the market sector as a whole (ABS 2016a).

The average age of Australia's aggregate capital stock in the market sector fell from 12.1 to 11.1 years between 2000 and 2015 (ABS 2015c). The market sector's capital-intensive industries, including mining, manufacturing, utilities and transport, carry high levels of capital stock reflected in high capital:labour ratios. In the construction sector, capital stock increased from AUD 43.7 billion to AUD 59 billion between 2010 and 2015 – an average growth rate of 7% per annum – and the average age of capital stock fell from 11.8 years to 11.1 years, reversing the ageing trend of the previous decade (ABS 2015c). However, the industry possesses the lowest value of capital stock per worker at AUD 142,857 in June 2015 compared with mining AUD 2.9 million, utilities AUD 2.5 million, media AUD 710,000 and transport AUD 660,000 (ABS 2015c). From this perspective, the construction industry possesses many of the characteristics of a service industry.

Technical progress

Endogenous growth theory recognises the importance of technical progress to sustain improvements in economic growth, productivity and per capita incomes over time. In practice, technical progress requires investment in research and development activity that develops the technologies that will improve capital productivity with flow-through effects to labour and multifactor productivity. Apart from research and development activity, technical progress generally includes investment in buildings, plant and equipment to harness embodied technologies, the rapid adoption of new technology acquired from other countries and investment in intellectual property. The boundaries are never closed, however, and technical progress can be achieved through investment in human capital, technology transfer and improvements in know-how by change in workplace management and practices.

Embodied technology

Investment in new plant and equipment introduces embodied technology, that is, improvements in the design, functionality or quality of new capital goods or intermediate inputs over time. Embodied technology is also described as the 'technological gap' and refers to the incremental productivity improvement of new plant and equipment compared with average and older machinery (Cummins and Violante 2002). Embodied technology may come into existence because the client has specified a higher level of performance, it may be mandated by regulation or it may be latent, as a result of the designer or manufacturer improving the specification, technology or performance of the latest models of capital goods in a competitive market environment. Cummins and Violante (2002) use a price-based approach to measure technology change at the asset, industry and aggregate level in the United States between 1947 and 2000. The industry-level findings suggest that technical change in equipment and software in this period was a significant factor in the productivity and growth resurgence of the 1990s.

Research and development

Research and development is systematic investigation and experimentation that leads to new knowledge and improved products, processes and services. Investment in research and development extends the technical frontier, contributes to growth in MFP (D'Arcy and Gustafsson 2012: 29) and is a long-term commitment on the part of firms that requires significant investment.

While Australia spends proportionally more on higher education research than the OECD average, business investment in research in the 10 years to 2015 was around 1.38% of GDP, lower than the OECD average of 2.27% (OECD 2015; Australian Government 2015: 48). The contribution of Australian businesses to collaborative research is also low, with 32% of large firms and 23% of small to medium firms entering into collaborative research agreements, compared with OECD averages of 55% and 24%, respectively (ABS 2015d). Very few business research agreements exist between universities and non-commercial research organisations, with 2% of large firms entering into collaborative research with this group of institutions, compared with an OECD average of 37%. In the UK, Canada and the US, nearly all research expenditure is funded with equity. In Australia, 71% of research is financed from borrowings, a short-term approach to research investment that is subject to early program termination or funding cuts that are a result of greater firm exposure to exogenous risk (Australian Government 2016).

In Australia, the construction industry accounted for around 9% of GVA and 10% of total employment, although only 1% of firms entered into medium-term research and development programs (ABS 2016b; Chubb 2013). An explanation for low participation in research and development may be the fragmentation of the industry, with 90% of registered firms employing fewer than five persons and only 3% of firms employing 1,000 or more persons (ABS 2016d; ONS 2016b). International data shows that most industry investment in research and development is made by large enterprises (Australian Government 2016). Construction accounts for 5% of business expenditure on research and development in Australia, which is low given the industry accounted for around 8.4% of GDP in 2014 (Australian Government 2015; ABS 2015a). The industry finances most of its research and development from current account, with less than 5% as capital expenditure (ABS 2015d).

A further explanation of the construction industry's low contribution to research may be volatility of earnings, which limits opportunities to all but the larger construction firms with continuity of work flows and those with sub-industry specialisations. Evidence from the UK Office of National Statistics shows that construction accounts for around 6% of GVA and 7% of employment in that country, yet construction and engineering firms account for less than 2% of applications for research and development tax incentives (ONS 2016b). The research and development challenge was identified as a problem for the industry by Egan (1998) and evidence suggests that little has changed.

Intellectual property

In OECD countries, investment in intellectual property is associated with economic growth, and in several European countries investment in intellectual property exceeds that for buildings, plant and equipment (Australian Government 2016). In Australia, most investment in intellectual property is provided by the leading knowledge-intensive industries of mining, manufacturing, media, and financial and professional services, with most investment in computer software and mineral and oil exploration. Following the winding-back of the mining boom, investment in computer software in non-mining industries increased, although using the long-term trend in GDP terms, the value has declined over the past decade.

Around 93% of Australian patent and innovation patent registrations are held by non-residents, with residents accounting for 9,497 patents in other countries, twice that of domestic registrations. Trademark applications were mainly made by resident firms and increased by 15% in 2015 over the preceding year, design right applications by 6% and plant breeders' rights by 12%. The majority of intellectual property applications over the past decade were made by small and medium-sized firms (33%), non-residents (52%), private owners (13%) and large enterprises (2%). The average trading history of applicants is seven or more years, with negligible applications from the construction industry. Australian firms make limited use of the patent and design registration system, with only 21% of enterprises protecting their intellectual property interests.

Vocational education and training

Vocational education and training (VET) improves the quality of human capital over time and, because of its positive effect on labour productivity, it is an important driver of multifactor productivity growth (ABS 2012; D'Arcy and Gustafsson 2012: 29). VET also plays an important role in improving the effectiveness of capital productivity and contributing to workplace innovation, research and development. In Australia VET is led by the knowledge-intensive industries, with 20% of firms reporting increases of 15% or more in expenditure between 2012 and 2016. Australian industry invests around AUD 10.6 billion annually in VET, with employers providing training for an average 32 hours per employee annually (Australian Government 2015: 120). At the industry level, construction firms provide less training time than all other industries, with an average 22 hours per person annually (Richardson 2004; NCVER 2015).

The favoured methods of firm training are informal training (77.6%), accredited training programs (48%) and non-accredited training (47.5%). The majority of this investment took place in the knowledge-intensive industries of mining, manufacturing, media, finance and insurance, and professional services. The construction sector, with its wide use of apprenticeships, was the biggest user of accredited training but was in the lowest percentile of industries supporting tertiary education programs for employees and collaborative research with tertiary institutions (NCVER 2015).

While VET has been shown to improve a nation's stock of human capital, the benefits to the firm may not materialise in the short to medium term. However, recruitment of personnel with proficiency in literacy, numeracy and information technology has been shown to improve short-term productivity (Brown *et al.* 2014). Literacy skills are associated with proficiency in problem-solving in technology-rich environments, which is linked to firm-level investment and improved productivity (Richardson 2004; OECD 2016b). Australia ranks in the top percentile of literacy skills in the OECD, with 38% assessed at levels two and three (OECD 2016b: 41). Australia ranks eighth in problem-solving in technology-rich environments with a score of 38 (OECD average 31) (OECD 2016b: 61). Globally, services to buildings ranks in the bottom five industries for reading, writing, numeracy, information technology and problem-solving skills (OECD 2016b: 109). This is consistent with Australian data showing fewer employees with high proficiency in literacy (less than 20% at levels two and three) and problem-solving skills (Brown *et al.* 2014).

Industry profitability

Measuring productivity at the enterprise level is problematic, with a number of factors contributing to year-on-year variations such as short-term changes in demand and supply and the business cycle. Long-term industry-specific shocks such as extended droughts, new technologies and change in terms of trade and exchange rates may affect industry productivity trends over time. A further factor is capacity utilisation, which may be affected by a downturn in the business cycle with further capital investment deferred until conditions point to a recovery in growth (Barnes 2011).

Investment in buildings, plant and equipment, employee training and research and development requires a significant number of firms to have the financial capacity to invest, which assumes a reasonable level of industry profitability. Profitability will determine whether firms will have the financial 'headroom' or operating surplus to meet bid costs for competitive tenders[6] and invest in measures designed to improve productivity after dividends and other calls on net operating profit have been met. The return on investment of Australian industries shows considerable differences, reflecting variation in industry-level operating conditions such as supply and demand, the level of competition, the price of inputs, the business cycle and, in the case of the transport and utilities industries, interventions by state regulatory bodies.

An analysis of the return on investment of selected market sector industries shows that the lightly capitalised construction industry had an average return on investment (return on assets/net capital stock) of 67.1% per annum between 2007 and 2015, the highest of all industries, compared with the retail industry (33.4%), wholesale (27.9%), manufacturing (17.3%) and mining (15.7%).[7] Data shows that the construction industry had a 10.25% average profit margin per annum that is partly attributable to a high revenue:capital stock ratio and the practice of accounting for major plant, equipment and site-specific costs as a project expense. In the 10 years to 2015, employee costs remained at 15%–17% of revenue and construction revenue grew at an annual average 7.1% per annum, suggesting that output, industrial relations and labour costs were not significant factors in the decline in industry MFP over this period. Using chain volume measures of output and GDP between 2007 and 2015, real growth in output in the mining industry, agriculture, the media and construction industries exceeded growth in GDP. However, profitability is a different question, and lower return on investment in the capital-intensive industries of transport, mining and utilities extended investment payback periods, slowed investment growth and contributed to an increase in the average age of capital stock between 2000 and 2015 in these industries (ABS 2015a).

At the firm level

Improved capital productivity at the firm level is essentially about improving output from given capital inputs through improved efficiency and organisation, investment in new buildings, plant and equipment, innovation and minimising waste. The construction industry faces a number of industry-specific challenges reforming business management practices with Australian firms scoring 2.98 (on a scale of 1 to 5) and ranking ninth in a sample of 21 countries surveyed by McKinsey and Company (2009). For the same survey in 2014, Australia scored 2.75, less than the mean of the sample of 20 countries (Dowdy and Van Reenen 2014). Other initiatives include improvements to workplace organisation, achieving economies of scale and ensuring stable revenue by adopting partnering and supply chain management methods to reduce risk and provide greater certainty with the cost of inputs. Many of the efficiency drivers that address capital deepening and MFP performance have been available for 20 years or more although culture change at the enterprise level, particularly in a highly fragmented industry in which the majority of firms employ five or fewer persons, can be a slow process.

In the construction industry, many initiatives aimed at improving performance have centred on time and cost performance (Egan 1998; National Audit Office 2001, 2003, 2005; Ernst and Young 2015). The specific enterprise reforms include wider use of partnering arrangements, improved risk identification and management practices, prefabrication and lean procurement, the use of non-traditional procurement methods and a supply chain approach to contracting relationships. Recent case studies have drawn on transactional evidence to identify opportunities to improve capital productivity in construction projects, with savings of more than 20% from the use of best practice capital productivity tools (Taylor *et al.* 2012: 33; Ernst & Young 2015).

Cost and schedule control

A significant and systemic problem in the construction industry is cost and time overruns, with cost overruns in building construction in the range 45%–73% and schedule overruns in the range 24%–70% (NAO 2003, 2005, 2009). The major reasons for cost and time overruns include poor project planning and opti-mism bias, risk management and an adversarial contracting environment (NAO 2005), a poorly informed client (Australian Government 2011) and failure to comply with the project evaluation and planning requirements in place at the time (QAO 2013). The problem is similar for projects in both the public and private sectors.

Innovation

Innovation in production processes and enterprise organisation are important in delivering new sources of growth and maintaining high-wage employment. Work-place innovation is a continuing process with two elements: workplace improve-ments designed to contain or reduce costs and investment in collaborative agreements with the nation's universities and public research institutions. Data shows that Australian firms invest less in collaborative research and development than the median for OECD countries (Australian Government 2016). Neverthe-less, opportunities exist for firm-level development of new fibre-reinforced build-ing materials, prefabricated low-cost housing, cost-effective trigeneration energy technology for commercial buildings, off-site prefabrication and sub-assembly, water recycling, improved energy efficiency for commercial and residential build-ings, and new construction methods that extend the life cycle of roads. Much of the technology is already in existence and requires funding to progress to the commercialisation stage of development (AIG 2014).

Improved efficiency

Innovative improvement in the workplace refers to the application of scientific approaches to workplace organisation and management. Efficiency may also be improved with the introduction of employee incentive programs that reward achievement of output and cost management objectives, changes in work prac-tices designed to improve operational efficiencies and cut costs, innovative pro-duction methods, cost-effective outsourcing of non-core services, encouragement of entrepreneurial initiatives and new business opportunities, and improving competitiveness through new ideas and technology. In the construction indus-try, considerable progress has been made over the past decade with the applica-tion of new contracting methods such as cost-led procurement and partnering arrangements with clients, sub-contractors and other stakeholders (UK Cabinet Office 2014; Egan 1998; NAO 2005), value for money tender evaluation prac-tices (Department of Treasury and Finance 2001), building information model-ling (Young *et al.* 2008), value engineering methods, lean construction (Howell 2013), life cycle costing (Regan *et al.* 2016) and deeper analysis of risk allocation in major projects (Department of Treasury and Finance 2003).

In a competitive bidding market, many of the benefits of the new methods accrue to clients rather than the construction industry, although cost-saving gains can now be shared between the client and contractors. Under these arrangements, contractors' liability for cost overruns is capped at firm overheads and profit margin under alliance and hybrid contracting arrangements. Progress has also been made in reducing bid costs using two-stage tendering.

Technical innovation, growth in capital productivity, capital deepening and improved procurement outcomes led to significant improvement in capital productivity in Australia between 2001 and 2014, with construction outperforming the market sector (ABS 2014b; Richardson 2014: 8). In the six years to 2014, construction also outperformed the market sector in labour productivity growth, with annualised growth of 6.38% for the building construction sub-sector and 3.51% for heavy and civil engineering. Annualised labour productivity growth for the construction industry averaged 4.81% over this period (ABS 2013).[8]

In the 10 years to 2015, the construction industry outperformed the market sector in measures of industry return on assets and labour and capital profitability while delivering output growth nearly double that of both market sector and economy-wide GDP growth. Why did this not transfer to better MFP? One reason may be the diminishing marginal return to capital inputs.

Capital productivity: the environmental factors

A number of systemic risks and uncertainties affect capital productivity performance. These include natural events such as adverse weather, fluctuating exchange rates and volatility in a country's macroeconomic variables including interest rates, fiscal policy, global trade and financial shocks. Activity levels in the construction industry are influenced by the business cycle, private investment and investor confidence, government spending programs and the property cycle. Construction activity may also be affected by mismatch between the business and investment cycles. Each of these factors is linked to the actions of public institutions and decision-making that further influences the business cycle, interest rates, private sector confidence, employment, future demand and activity levels.

Public institutions

Public institutions also determine the incentives and constraints on the actions of parties to commercial transactions (Acemoglu *et al.* 2004; Williamson 1985). They directly affect capital investment by industry through industry policy, trade protection measures, the creation of tax incentives such as accelerated depreciation, special capital allowances, tax exemptions and offsets. Tax concessions may be generic in scope, as occurs in Singapore with 'pioneer status' capital allowances for investment in intellectual property assets, or in the United States with the use of tax-effective enterprise structures to encourage private investment in emerging technology industries.

The effectiveness of state institutions in attracting private investment in industries and projects was demonstrated in China during the 1980s when the government decided to fast-track development of its telecommunications sector with a series of institutional changes and initiatives designed to create a favourable environment for private investment in the sector (Lu 2000: 385). Between 1984 and 1995, the effective life of telecommunications equipment was shortened from 20 years to 5 years; preferential import tariffs for telecommunications equipment were initially lowered and then eventually abolished for foreign investment, while reductions were made in sales taxes. In the 10 years to 1995, investment in the telecommunications sector increased threefold, with most new investment provided by private firms (Lu 2000: 388). The approach was subsequently applied to a number of priority industries in early 1991 and a zero tax rate applied for prioritised projects delivering transport infrastructure and utility services.

Government capital spending

Government capital spending, particularly in capital-intensive networked assets, has been shown to create a co-integration relationship between public and private capital, giving rise to a 'crowding-in' effect with private investment (Abiad *et al.* 2015; Economic Insights 2015; Hatano 2008). The evidence suggests a long-term multiplier effect flowing from public to private capital up to 1.6 times the level of public investment with a positive and direct association to productivity (Isaksson 2007; Destefanis and Sena 2005). In the United States, the Regional Input-Output Modelling System (RIMS II) was introduced by the Bureau of Economic Analysis (BEA) in 2007 to calculate the annual regional multipliers for output, earnings, employment and value added for various industries. Multipliers measure the incremental economic activity that flows from construction spending. The industry was found to generate 1.9 times additional economic activity during construction and around 1.4 times post-construction (BEA 2010).[9] Recent studies that examined the effects of public investment in capital-intensive assets such as infrastructure found multipliers exceeding those of other activities (Leduc and Wilson 2012).

Externalities

The efficiency of firm investment in capital and labour productivity may be affected by externalities or events beyond the control of management. Several of these have their source in natural phenomena such as storms, unseasonal variations, adverse weather and other natural events, and environmental factors. Other determinants may be the result of government interventions such as the actions of regulatory agencies, new laws and compliance regimes, the speed and cost of transactions with government and the adequacy of public infrastructure including communications and information networks and systems. A further category of influences beyond the control of enterprise managers may be market driven; they include commodity prices, availability of building materials, the terms of trade, economic shocks (such as the global financial crisis in 2008), the business cycle and level of demand in the economy. These factors impact the pricing of capital and labour inputs and, particularly, the value of output. Externalities affect productivity performance in the construction industry although capital productivity investment, management and, ultimately, performance are in the hands of firms (Dowdy and Van Reenen 2014; Solow 2014).

Conclusion

A long-term view of productivity performance in the market sector in Australia shows annual growth between 1974 and 2015 of 3% in output, 2.2% in inputs and 0.8% in MFP.[10] To achieve this, capital inputs averaged 4.4% and labour inputs 0.7% with substantial improvement in capital deepening over this period. A more recent view of productivity performance indicates that MFP growth was negative in the period 2013–2016, which cannot be explained by resource real-location or the terms of trade. The data also points to differences at the industry level, with traditional drivers of GVA showing growth in capital productivity offset by lower growth in outputs.

Capital productivity is an important driver of growth in the economy and relies on firm investment in new plant and equipment, research and development, edu-cation and training and workplace innovation. However, methods of estimation prior to 2015 were unreliable because of the limitations of the Solow growth accounting framework, reliance on a small number of proxy indicators, inputs and pricing techniques that encounter heterogeneity problems, the exclusion of intermediate inputs and inconsistencies in national measurement. The recent adoption of the KLEMS system is expected to deliver more accurate estimates, but limitations still exist. The Australian construction industry shares many of the characteristics of the industry in OECD countries with low levels of invest-ment in intellectual property, collaboration, new contracting methods, research and development and human capital. The trend suggests that the industry world-wide is underinvesting to meet the commercial and technological opportunities and challenges of the next 25 years.

Capital productivity in the construction industry between 2004 and 2016 was strong, albeit from a low base. The construction industry has experienced less capital deepening than other industries and MFP performance has been adversely affected by modest improvement in labour productivity. The construction indus-try has a poor record of investment in tangible assets and human capital, and its workforce is the lowest in the market sector for education and training. The reasons for low industry capital productivity between 2010 and 2016 is partly explained by industry dynamics including fragmentation over a large number of small firms, tendering practices, low levels of industry and research collabora-tion and the adversarial nature of tendering processes and contracts which lim-its opportunity for wider use of alliance contracting and cost-plus procurement methods that are widely used in North America and Europe. The industry is characterised by short-term employment contracts, wide use of sub-contractors and an unwillingness to plan much beyond the short term. Initiatives were intro-duced in the UK following the Egan Report and some progress is being achieved with government support for the use of supply chain and partnering approaches, shared incentives, risk sharing and hybrid contracting forms for public contracts. In Australia, the cultural change towards new ways of conducting business in the industry has been a slow process.

To improve capital productivity performance in Australia, most of the challenges exist at the firm level. Improved management efficiency, investment in physical and human capital, and workplace innovation are specific initiatives in which the Australian construction industry trails OECD averages. Robert Solow, the architect of the neoclassical growth model, argues that in the present environment, improved capital productivity at the firm level is most likely to be achieved by firms exercising quality decision-making and competing in a market with others exercising world's best practice (McKinsey 2014).

Notes

1 In Australia, labour productivity is generally expressed as output per hour worked, although other measures are used by OECD countries (ONS 2016b).
2 KLEMS stands for capital (K), labour (L), energy (E), materials (M) and service (S) inputs.
3 Growth peaks are used because they represent similar levels of capacity utilisation in the economy. Productivity estimates are taken during productivity cycles at the peak in the cycle (when capacity utilisation is highest). At no other point in the cycle can we say that this occurs, which makes it hard to measure performance other than at the peak.
4 The neoclassical growth accounting framework was originally adopted by the OECD to calculate productivity and provide government with the policy levers to encourage improvement in MFP performance. More recent work in the mid-1980s by Romer (1987), Lucas (1988) and Rebelo (1991) established endogenous growth theory. Central to this new work is the proposition that growth may continue indefinitely because the returns to investment in a broad class of capital goods (including human capital) do not diminish as economies develop (Barro and Sala-I-Martin 2003:20). This is negated by spillovers of knowledge across producers and the externalities generated by advancing human capital (especially health and education). Romer (1987, 1990), Aghion and Howitt (1992) and Grossman and Helpman (1991) incorporated research and development and competition into the endogenous growth framework. They argued that technological advance results from purposive research and development activity that creates a form of *ex post* monopoly power. As long as ideas and inventions flourish, the long-run growth rate can remain positive (Regan 2007).
5 The Australian Bureau of Statistics calculates capital stock from expenditure on non-dwelling buildings, machinery and equipment, weapons systems, cultivated biological resources, research and development, mineral and petroleum expenditure and computer software (ABS National Accounts 2014a).
6 Bid costs average around 1% of contract value for conventional construction projects valued at AUD 50 million or more, and up to 5% for megaprojects valued above AUD 500 million (Regan et al. 2015).
7 Return on investment is calculated using average annual earnings before taxation divided by year-end net capital stock (ABS 2014b, 2015c).
8 Heavy and civil engineering construction accounted for 42% of industry GVA in 2013, building construction 34% and construction services 24% (ABS 2013).
9 Multipliers recognise two stages in the production process. For buildings this means the initial investment stage (jobs, employee spending, payments to suppliers and contractors) and the post-construction stage (rent income, expenses, management fees,

utilities and rates, and life cycle repairs and maintenance costs). Economic multipliers are published in OECD countries for all industrial classifications as a standard measure of economic activity.

10 It is worth noting that MFP went down despite significant growth in capital productivity. Without the contribution of capital productivity, the result would have been negative, placing Australia at the bottom of OECD rankings. The problem is labour productivity, even after allowing for capital deepening. It is worth noting that productivity estimates for all OECD countries are broad estimates and the methodology used has many shortcomings.

References and further reading

Abiad, A., Furceri, D. and Topalova, P. (2015) *The Macroeconomic Effects of Public Investment: Evidence from Advanced Economies*, Working Paper WP/15/95, Research Department (Washington, DC: International Monetary Fund).

ABS (2007) *Experimental Estimates of Industry Multifactor Productivity*, Information Paper, Appendix 2 Sensitivity Analysis of Capital Inputs, Cat. No. 5260.0.55.001 (Canberra: Australian Bureau of Statistics).

ABS (2012) *Measures of Australia's Progress: Summary Indicators*, Cat. No. 1370.0.55.001, October (Canberra: Australian Bureau of Statistics).

ABS (2013) *Australian Industry 2012–13*, Cat. No. 8155.0, May (Canberra: Australian Bureau of Statistics).

ABS (2014a) *Australian System of National Accounts, 2012–13*, Cat. No. 1370.0.55.001, October (Canberra: Australian Bureau of Statistics).

ABS (2014b) *Estimates of Industry Multifactor Productivity*, 2013–14, Cat. No. 5260.0.55.002, December (Canberra: Australian Bureau of Statistics).

ABS (2014c) *Research and Development by Private Firms 2013–14*, Cat. No. 8104.0 (Canberra: Australian Bureau of Statistics).

ABS (2015a) *Australian Industry 2014–15*, Cat. No. 8155.0, May (Canberra: Australian Bureau of Statistics).

ABS (2015b) *Productivity in the Market Sector, Growth Cycle Analysis*, Cat. No. 5260.0, Table 14, October (Canberra: Australian Bureau of Statistics).

ABS (2015c) *Australian System of National Accounts, 2013–14*, Cat. No. 5204.0, October (Canberra: Australian Bureau of Statistics).

ABS (2015d) *Research and Experimental Development, Business*, Australia 2013–14, Cat. No. 8104.0, May (Canberra: Australian Bureau of Statistics).

ABS (2015e) *Experimental Estimates of Industry Level KLEMS Multifactor Productivity*, Information Paper, Cat. No. 5260.0 (Canberra: Australian Bureau of Statistics).

ABS (2016a) *Australian System of National Accounts, 2014–15*, Cat. No. 5204.0, October (Canberra: Australian Bureau of Statistics).

ABS (2016b) *Labour Force Australia Quarterly May 2016*, Cat. No. 6291.0.55.003, June (Canberra: Australian Bureau of Statistics).

ABS (2016c) *Engineering Construction Activity*, Cat. No. 8762.0, June (Canberra: Australian Bureau of Statistics).

ABS (2016d) *Australian Industry*, Cat. No. 8155.0, May (Canberra: Australian Bureau of Statistics).

Access Economics and World Competitive Practices Pty. Ltd. (1999) *Australian Construction Productivity: International Comparison*, Report prepared for Australian Constructors Association, August (Canberra).

Acemoglu, D., Johnson, S. and Robinson, J. A. (2004) *Institutions as the Fundamental Cause of Long-Term Growth*. National Bureau of Economic Research, Working Paper 10481, May.

Aghion, P. and Howitt, P. (1992) A model of growth through creative destruction. *Econometrica*, **60**, 322–351.

AIG (2014) *Boosting the Commercial Returns from Research*, Consultation Paper, November (Sydney: Australian Industry Group). 13 August 2016. cdn.aigroup.com.au/Submissions/General/2014/Ai_Group_Boosting_the_commercial_returns_from_research_submission_November_2014.pdf.

Australian Government (2011) *Building the Education Revolution, Implementation Task Force*. Final Report, Canberra, July.

Australian Government (2015) *Australian Innovation System Report*. Office of the Chief Economist, Department of Industry, Innovation and Science, Canberra.

Australian Government (2016) *Industry Monitor*. Office of the Chief Economist, Department of Industry, Innovation and Science, Canberra.

Australian National Audit Office (2010) *Major Projects Report, Defence Materiel Organisation*. Assurance Report No. 17, Canberra.

Barnes, P. (2011) *Multifactor Productivity Growth Cycles at the Industry Level*. Productivity Commission Staff Working Paper, Canberra.

Barro, R. and Sala-i-Martin, X. (2003) *Economic Growth*, 3rd ed. (New York: McGraw-Hill).

Brown, J., McKenzie, P. and Taylor, M. (2014) *The Returns on Investing in the Workplace*. Presentation at the National Adult Language, Literacy and Numeracy Conference held in Melbourne, Australian Council for Education Research-Australian Industry Group, Melbourne.

Bureau of Economic Analysis (2010) *Input-Output Account Analysis*. www.bea.gov/industry/io_annual.htm.

Choi, J. and Russell, J. (2004) Economic gains around mergers and acquisitions in the construction industry in the United States of America. *Canadian Journal of Civil Engineering*, **31** (3), 513–525.

Chubb, I. (2013) *The Impact of Research and Development in Construction*. Address to the Committee for the Economic Development of Australia, Brisbane 7 May, Australian Government, Canberra.

Cummins, J. G. and Violante, G. L. (2002) Investment-specific technical change in the US (1947–2000): Measurement and macroeconomic consequences. *Review of Economic Dynamics*, **5** (2), January, 243–284.

D'Arcy, P. and Gustafsson, L. (2012) Australia's productivity performance. *Reserve Bank of Australia Bulletin*, June Quarter, 23–35.

Deloitte Access Economics (2014) *Victorian Construction – Labour Costs and Productivity*. Master Builders Association of Victoria, Final Report, December.

DeLong, J. B. and Summers, L. H. (1991) Equipment investment and economic growth. *The Quarterly Journal of Economics*, **106** (2), 445–502.

Department of Treasury and Finance (Victoria) (2001) *Practitioners Guide*. Partnerships Victoria Guidance Materials, Melbourne.

Department of Treasury and Finance (Victoria) (2003) *Public Sector Comparator, Supplementary Technical Note*. Partnerships Victoria Guidance Materials, Melbourne.

Destefanis, S. and Sena, V. (2005) Public capital and total factor productivity: New evidence from the Italian regions, 1970–98. *Regional Studies*, **39** (5), 603–617.

Diewert, W. E. and Fox, K. J. (2008) On the estimation of returns to scale, technical progress and monopolistic markups. *Journal of Econometrics*, **145** (1–2), 174–193.

Domar, E. D. (1946) Capital expansion, rate of growth, and employment. *Econometrica*, **14** (2), 137–147.

Dowdy, J. and Van Reenen, J. (2014) Why management matters for productivity. *McKinsey Quarterly*, September.

Dozzi, S. P. and AbouRizk, S. M. (1993) *Productivity in Construction*. Institute for Research in Construction, National Research Council, Ontario.

Economic Insights Limited (2015) *What Is the Relationship Between Public and Private Investment in Science, Research and Innovation?* A Report commissioned by the Department for Business, Innovation and Skills, London, April.

Edwards, J. (2014) *Beyond the Boom*. (Melbourne: Penguin Group).

Egan, J. (1998) *Rethinking Construction*, Report of the Construction Task Force for the Deputy Prime Minister John Prescott, Department of Trade and Industry (Norwich: HMSO).

Ernst and Young. (2015) *Opportunities to Enhance Capital Productivity*, a report by the EY Global Mining and Metals Network (London: EY Global Limited).

Eslake, S. and Walsh, M. (2011) *Australia's Productivity Challenge* (Melbourne: Grattan Institute).

EU (n.d.) *EU KLEMS: Capital, labour, energy, materials and service. Eurpoea Commission.* https://ec.europa.eu/info/business-economy-euro/indicators-statistics/economic-databases/eu-klems-capital-labour-energy-materials-and-service_en

Foster, J. (2015) The Australian multi-factor productivity growth illusion. *Australian Economic Review*, **48** (1), 33–42.

Furneaux, C., Brown, K. and Hampson, K. (2007) Mapping the Australian regulatory environment: Implications for construction firms. In: *Proceedings of the International Council for Research and Innovation in Building and Construction*, World Building Congress 2007, Paper CIB2007–095.

Grossman, G. M. and Helpman, E. (1991) *Innovation and Growth in the Global Economy* (London: The MIT Press).

Harrod, R. F. (1939) An essay in dynamic theory. *The Economic Journal*, **49** (193), 14–33.

Hatano, T. (2008) *Crowding-in Effect of Public Capital on Private Capital*. Public Capital, Vol. 89. Policy Research Institute, Ministry of Finance, Government of Japan.

Howell, J. (2013) Lean construction. *Public Infrastructure Bulletin*, October (9), 34–43.

Isaksson, A. (2007) *Determinants of Total Factor Productivity: A Literature Review*, Staff Working Paper 02/2007, Research and Statistics Branch, United Nations Industrial Development Organisation, Vienna.

Infrastructure Partnerships Australia (2007) *Performance of PPPs and Traditional Procurement in Australia*. Sydney. http://infrastructure.org.au/wp-content/uploads/2016/12/IPA_PPP_FINAL.pdf

Jorgenson, D. W., Ho, M. S. and Stiroh, K. J. (2007) *A Retrospective Look at the U.S. Productivity Growth Resurgence*. Staff Report No. 277, Federal Reserve Bank of New York, February.

Kent, C. (2014) *Non-Mining Business Investment – Where to from here?* Address to the Bloomberg Economic Summit held in Sydney on 16 September 2014. www.rba.gov.au/speeches/2014/sp-ag-160914.html.

KPMG (2016) *The Role of Capital and Labour in Driving Economic Growth*. Research Paper, KPMG Economics, February.

Lane, K. and Rosewall, T. (2015) *Firm-Level Capacity Utilisation and the Implications for Investment, Labour and Prices*, Bulletin, December Quarter (Sydney: Reserve Bank of Australia). www.rba.gov.au/publications/bulletin/2015/dec/pdf/bu-1215-2.pdf.

Latham, M. (1994) *Constructing the Team*, Final Report of the Government and Industry Review of Procurement and Contractual Arrangements in the UK Construction Industry (London: HMSO).

Leduc, S. and Wilson, D. (2012) *Roads to Prosperity or Bridges to Nowhere? Theory and Evidence on the Impact of Public Infrastructure Investment*. Working Paper No. 18042, National Bureau of Economic Research, Washington, May.

Lu, D. (2000) China's telecommunications buildup: On its own way. In: Ito, T. and Krueger, A. O. (eds.) *Deregulation and Interdependence in the Asia-Pacific Region* (Chicago: University of Chicago Press).

Lucas, R. E. (1988) On the mechanics of economic development. *Journal of Monetary Economics*, **22**, 3–42.

McKinsey and Company. (2009) *Management Matters* (Washington).

McKinsey and Company. (2014) Prospects for growth: An interview with Robert Solow. *McKinsey Quarterly*, September.

NAB. (2016) *NAB Monthly Business Survey*. NAB Group Economics, January (Melbourne: National Australia Bank).

NAO (2001) *Modernising Construction*. Report by the Comptroller and Auditor General, HC87 Session 2000–2001, 11 January. National Audit Office.

NAO (2003) *PFI: Construction Performance*. Report by the Comptroller and Auditor General, HC 371 Session 2002–03, 5 February. National Audit Office.

NAO (2005) *Improving Public Services Through Better Construction*. Report by the Comptroller and Auditor-General, HC364–1 Session 2004–05, 15 March.

NAO (2009) *Performance of PFI Construction*. A review by the Private Finance Practice, October, London.

NAO (2010) *The Major Projects Report*. Report by the Comptroller and Auditor General, HC489–1 2010–11, 15 October.

NCVER (2015) *Survey of Employer Use and Views, Training Choices 2005–2015*. National Centre for Vocational Education Research (Canberra: Australian Government).

NZIER (2013) *Construction Industry Study, Implications for Cost Escalation in Road Building, Maintenance and Operation*. New Zealand Institute of Economic Research, Final report to the Ministry of Transport, November.

OECD (2003) *The Sources of Economic Growth in OECD Countries* (Paris: OECD).

OECD (2015) *Main Science and Technology Indicators Database* (MTSI 2014/2). www.oecd.org/sti/msti.htm.

OECD (2016a) *Compendium of Productivity Indicators* (Paris: OECD).

OECD (2016b) *Skills Matter, Further Results from the Survey of Adult Skills* (Paris: OECD).

ONS (2016a) *Blue Book*. Office of National Statistics (UK). www.ons.gov.uk/releases/uknationalaccountsthebluebook 2016.

ONS (2016b) *International Comparisons of Productivity: Final Estimates 2014* (London: Office of National Statistics).

Parham, D. (2013) Australia's productivity: Past, present and future. *Australian Economic Review*, **46** (4), 462–472.

Parliament of the Commonwealth of Australia (2010) *Inquiry into Raising the Productivity Growth Rate in the Australian Economy*. House of Representatives, House Standing Committee on Economics, Final Report, Canberra.

Pereira, A. and Andraz, J. M. (2013) *On the Economic Effects of Public Infrastructure Investment: A Survey of the International Evidence*. Working Paper #108, Department of Economics, College of William and Mary, Virginia.

Pereira, A. and Roca-Sagales, O. (1999) Public capital formation and regional development in Spain. *Review of Development Economics*, **3** (3), 281–294.

Productivity Commission. (1991) *Construction Costs of Major Projects*. Industry Commission Report No. 8 (Canberra: Australian Government Publishing Service).

Productivity Commission (2013) *PC Productivity Update 2012*, May (Canberra: Australian Government Publishing Service).

Productivity Commission (2014a) *Public Infrastructure*. Inquiry Report No. 71 (Canberra: Australian Government Publishing Service).

Productivity Commission (2014b) *PC Productivity Update 2013*, May (Canberra: Australian Government Publishing Service).

Productivity Commission (2015) *PC Productivity Update 2014*, May (Canberra: Australian Government Publishing Service).

Productivity Commission (2016) *PC Productivity Update*, April (Canberra: Australian Government Publishing Service).

PWC (2016) Global engineering and construction M&A deals. *Insights*, Quarter 2. PricewaterhouseCoopers.

Queensland Audit Office (2014) *Hospital Infrastructure Projects*. Final Report (Brisbane: Queensland Audit Office).

Rebelo, S. (1991) Long run policy analysis and long run growth. *Journal of Political Economy*, **99**, 500–521.

Regan, M. (2007) *Productivity and its Measurement, A Literature Review*, Working Paper 2006–11 (Hawthorn: School of Enterprise, University of Melbourne).

Regan, M., Smith, J. and Love, P. E. D. (2015) Public infrastructure procurement: A review of adversarial and non-adversarial contracting methods. *Journal of Project Procurement*, **15** (4), 425–457.

Regan, M., Smith, J. and Love, P. E. D. (2016) Whole life costing of infrastructure investment: Economic and social infrastructure projects in Australia. In: Mair, R. J., Soga, K., Jin, Y., Parlikad, A. K. and Schooling, J. M. (eds.) *Transforming the Future of Infrastructure through Smarter Information*, Proceedings of the International Conference on Smart Infrastructure and Construction, Cambridge University, 27–29 June (London: Institute of Civil Engineers (ICE) Publishing).

Romer, P. (1987) Crazy explanations for the productivity slowdown. In: Fisher S. (ed.) *National Bureau of Economics Research Macroeconomics Annual* (Cambridge, MA: The MIT Press).

Romer, P. (1990) Endogenous technical change. *Journal of Political Economy*, **98**, 71–102.

Richardson, D. (2014) *Productivity in the Construction Industry*. Technical Brief No. 33 (Canberra: The Australia Institute).

Richardson, S. (2004) *Employer's Contribution to Training*. National Centre for Vocational and Educational Research, Adelaide.

Runeson, G. and de Valance, G. (2009) The new construction industry. In: Ruddock, L. (ed.) *Economics for the Modern Built Environment* (London: Routledge).

Smout, M. (2014) *Why Are so Few Research and Development Tax Credit Claims from Engineering and Construction Firms?* ForrestBrown, UK. https://forrestbrown.co.uk/news/why-are-there-so-few-research-and-development-tax-credit-claims-amonFiggst-engineering-and-construction-businesses/.

Solow, R. M. (1956) A contribution to the theory of economic growth. *Quarterly Journal of Economics*, **50**, 65–94.

Solow, R. M. (2014) Prospects for growth. *McKinsey Quarterly*, September. www.mckinsey.com/global-themes/employment-and-growth/prospects-for-growth-an-interview-with-robert-solow.

Stiroh, K. J. (2001) What drives productivity growth? *Economic Policy Review*, Federal Reserve Bank of New York, March.

Taylor, C., Bradley, C., Dobbs, R., Thompson, F. and Clifton, D. (2012) *Beyond the Boom: Australia's Productivity Imperative* (New York: McKinsey Global Institute).

UK Cabinet Office (2014) *Cost Led Procurement Guidance*, Guidance for the procurement and management of capital projects (London: Controller of HMSO).

Williamson, O. E. (1985) *The Economic Institutions of Capitalism* (New York: The Free Press).

Young, N. W., Jones, S. A. and Bernstein, H. M. (2008) *Building Information Modelling, Transforming Design and Construction to Achieve Greater Industry Productivity* (New York: McGraw-Hill, Smart Market Report).

Editorial comment

When the Punjab region of India lost its capital, Lahore, after the partition of India and Pakistan in 1947, India's first prime minister, Jawaharlal Nehru, envisioned a new capital city for the region. The Swiss-French architect known as Le Corbusier completed the city plan and designed a number of unique public buildings located in the Capitol Complex, which was listed as World Heritage by UNESCO in 2016. Those buildings are constructed largely of raw reinforced concrete with bold off-form finishes, the so-called *béton brut* that was synonymous with much of Le Corbusier's work.

Little in the city plan or the major buildings is typically Indian; rather, the architect brought European ideas and construction technology to the north of India and inserted them into a somewhat hostile natural environment. Temperatures in Chandigarh can reach 45°C in the summer and fall to as low as 0°C during winter; average annual rainfall is close to 1,000 mm. It is arguable that the use of essentially European methods in this environment was not ideal, as the fabric of the buildings has deteriorated and extensive restoration work has been required. Ongoing maintenance is needed to combat the effects of climate and heavy use of air conditioning.

At the same time that Le Corbusier was working in Chandigarh, another European architect, British-born Laurie Baker, was working in India. He followed a very different approach, investigating regional building practices and using local materials. The question can be asked: what prompts the use of exotic materials and techniques in such places, rather than using Baker's approach and working with what is available and well understood by the people who will build, use and maintain the resultant structures?

In the following chapter, the author employs an innovative method to explore relationships between labour and materials costs in various countries in an attempt to identify the most logical choices for materials, construction methods and capital input into construction. The premise for the work is that it is not uncommon for lower-income (developing) countries to seek to use construction technologies that are not the most appropriate for their circumstances, which often include limited availability of modern materials that are not produced locally (and are thus imported and expensive) and require sophisticated equipment for installation (and perhaps ongoing maintenance) when local labour is plentiful and cheap.

8   Cost ratios and technology choice

Toong Khuan Chan

Introduction

This chapter explores the use of construction labour, material and plant cost ratios to develop a framework for evaluating the choice of construction materials and technologies. An important consideration for the selection of materials is often availability, such as locally grown timber or stone quarried from a local source. Additionally, the skills of local workers are very often closely matched with the type of material that is most commonly available, and thus building solutions have often relied on access to local materials and indigenous skills. However, modern construction methods now offer the builder many options, with a wider selection of materials together with a range of construction processes, some of which may be procured from overseas. Tatum (1988) defined this combination of resources, processes and conditions that produce a construction product as construction technology.

Fundamentally, the total cost of construction will comprise the cost of the raw materials; the labour to shape and assemble these materials; the purchase (or rental) of tools, machinery and other construction equipment; overheads (site-related costs, management, head office, compliance with all regulations, fees and insurances); and finally the builder's profit. Hence to reduce the overall construction or total life cycle cost of buildings, builders or their clients or designers may choose to adopt local construction methods, and materials that are durable and inexpensive to maintain. Modern construction methods employ various elements or sub-assemblies that can be fabricated elsewhere to reduce the physical work at the construction site. For example, prefabricated roof trusses, wall frames, window and door sub-assemblies are now ubiquitous in residential buildings.

Citing the gains in productivity in the manufacturing industry, Warszawski (1999) suggested that radical improvement of productivity and quality in building construction can be attained only through intensive industrialisation and automation of the building process. This means that many developing economies, faced with increasing demand for building products and services, are finding it difficult to develop policies that advance their local construction industries in the most appropriate directions with regards to construction materials and technologies. A careful choice of both material and technology can have significant economic and social consequences.

When considering projects in different international locations, construction clients are keen for advice about the total cost of projects at the feasibility stage and prior to bidding and construction. It follows that consultants therefore require information on material costs and the availability of plant and equipment to help them propose suitable construction technologies for these projects. Knowledge of the variance in costs of materials and technology can thus help all parties to take advantage of opportunities to use an efficient and economical combination of widely available materials and appropriate construction technologies for their project.

The aim of this chapter is to examine the drivers and factors relevant to the selection of materials and technology for the construction of buildings in several countries around the world. A series of construction cost ratios is computed based on selected basic inputs necessary for the construction of buildings. Differences in local practice, availability of local resources (raw materials, labour, capital and technology), domestic building materials industries and local regulations all combine to influence the construction cost of a building. The ratios are derived from a compilation of the costs of building material, workers' wages and the unit rates for various building elements from an international cost database. In this chapter, decisions on appropriate materials and technology are examined by comparing the cost ratios for labour, materials and plant to uncover patterns of use in 27 countries. The findings will inform current research and policy initiatives to manage the exploitation of indigenous resources, to develop domestic building materials industries, to improve construction methods and to modernise and upgrade the construction sector, particularly in developing countries. A rational method for selecting an 'appropriate' building technology to suit the conditions of the construction sector in different economies is invaluable, as policymakers seek solutions that will make construction work affordable and pursue policies that encourage the development of their construction sectors.

Cost comparisons and technology options

Previous work in this area of research can be traced to three lines of enquiry: the neoclassical macroeconomic theory of production by Cobb and Douglas (1928) that was first to quantify the relationship that existed between labour and capital in manufacturing production; the choice of an 'appropriate technology' for developing countries by Hillebrandt (1974), Moavenzadeh and Rossow (1975), Ganesan (1979), Gaude and Watzlawick (1992) and the International Labour Office (ILO 1998); and the more recent efforts by construction experts in the refinement of construction price surveys across a number of countries including efforts to calculate purchasing power parities (PPPs) for the World Bank's International Comparison Program (ICP).

The Cobb-Douglas production function which quantifies output as a function of labour L and capital C is

$$P = bL^kC^{1-k} \qquad\qquad\qquad\qquad (\text{Eq. 8.1})$$

where k and $1 - k$ were found using the method of least squares to be 0.75 and 0.25, respectively, and b was 1.01. In simple terms, Equation 8.1 states that a 1% increase in capital usage would lead to an approximate 0.25% increase in output while a similar 1% increase in labour would lead to a 0.75% increase in output. The production function has constant returns to scale, meaning that doubling the usage of both capital C and labour L will also double output P. This production function was the first to estimate the contributions of labour input and capital input to the value of all goods produced in a year. Derived primarily from aggregated industry output in the United States from 1899 to 1922, the equation has been used to describe the potential substitution of labour and capital based on the relative prices of these factors of production. Since its introduction in 1928, detractors have argued that this model has no basis or relationship to the technology or management of the production process and is merely a statistical correlation (Douglas 1967). For example, Robinson (1953) questioned the definition of capital when attempting to quantify the capital input into the production function. Notwithstanding numerous deficiencies, Equation 8.1 can be construed as one of the earliest attempts to demonstrate technology choice, albeit for the national economy.

The choice between utilising more labour or more plant and equipment (i.e. capital) in construction was discussed by Hillebrandt (1974). The choice of materials used can directly affect the method of construction, as can the availability and cost of the inputs, including the labour and plant and machinery required for construction. All technically possible combinations of capital and labour that can produce a fixed number of units of output were demonstrated using a series of iso-product (also known as *isoquant* in microeconomic theory) curves. As these iso-product curves were asymptotic to both the capital and labour axes, increasing amounts of capital were required with each unit reduction in manpower in the substitution of capital for units of labour (and vice versa), demonstrating an increasing rate of marginal substitution. When the relative prices of inputs change, a new equilibrium point can be established with a different capital/labour ratio compared to the previous prices. This goes some way to explaining the substitution of scarce or expensive labour with plant and equipment (i.e. capital inputs) in many developed economies.

Multilateral agencies such as the ILO (ILO 1998, 2006) have supported the use of more labour-intensive construction technologies in developing countries to (1) address the issue of unemployment or under-employment, (2) encourage the use of indigenous materials that need little plant or equipment to further process and (3) reduce the outflow of hard currency to purchase expensive materials or technologies from more advanced economies. In such cases, there are cost advantages in using labour-intensive technologies. There are, however, several disadvantages, such as the risk of not being able to upgrade the skills of workers or to produce a more sophisticated product based on the use of indigenous materials and technology.

The use of labour-intensive methods of construction also limits the ability of local contractors to compete against multi-national contractors. Moavenzadeh and Rossow (1975) suggested that multi-national construction firms may transfer, adapt and develop appropriate technologies to less industrialised countries to stimulate economic development, employment creation and income generation. They argue that there should be incentives to encourage multi-national construction firms to enter into joint ventures or sub-contracting arrangements with local firms to promote the use and training of local manpower, to develop middle management at the local level and also to use technologies appropriate to local conditions.

Hillebrandt (1999) attempted to address the dichotomy between job creation and the use of equipment by identifying the factors that may determine the appropriate balance in various countries. The argument for more capital-intensive technologies lies where a country has a high level of gross per capita income and low unemployment. The use of labour-intensive technology in this case can lead to wage inflation. There is clearly a need to consider a range of factors when deciding on the policy of 'appropriate technology', where some parts of the economy in a developing country may choose to adopt labour-intensive technologies while other parts adopt sophisticated technologies to boost productivity, develop capacity of the local industry to undertake more complex projects, and to eventually gain the opportunity to export their construction services.

The ILO (2001) reported that in countries where wages are low and there is mass unemployment, the replacement of labour by machines does not make sense, from either an economic or a social perspective. It has long been an approach by the ILO to promote the technical feasibility and economic viability of more labour-intensive methods of construction and the opportunity it affords for worker upskilling. Construction is often one of the few alternatives to agricultural labour for those who do not have any particular skill. A shortage of skilled labour is a factor behind the drive in many developed countries to mechanise production in order to raise productivity by replacing labour with plant and equipment. This makes sense only where full employment is creating upward pressure on wages but where there is surplus labour and high unemployment, as in the case of many lower-income countries, where employment in construction needs to be expanded to help absorb the growing labour force and to lift people out of poverty.

In the case of Sri Lanka, Ganesan (1994, 2000) clearly advocated a strategy to maximise construction sector employment given the country's labour surplus and limited foreign resources. Increased costs are incurred when international contractors propose more sophisticated technologies combined with high overhead costs plus salaries of expatriate managers. These cost premiums may be appropriate for complex projects such as large power generation plants, but often costs can be lower by using local technologies and domestic builders for less complex projects. The dependence on imports of construction components and materials combined with issues of high unemployment can lead to high construction costs, inflation and an unstable economy in some developing countries (Sultan and Kajewski 2006). Thus, policies put forward by various countries to improve the economic performance of their respective construction industries need to be informed by a realistic economic model that illustrates the link between the cost of inputs to the construction industry to the price of its outputs and related benefits to the national economy.

These prior investigations into the economics of the substitution of capital for labour and the arguments for labour-intensive technologies for employment maximisation have guided the research described here. However, comparing the cost of labour, materials or equipment across countries using exchange rates can often be distorted by the large differences in price levels between developed and developing countries. The representation of factor inputs as ratios, which eliminates the need for exchange rate conversions or PPP adjustments, is described in the next section.

Cost studies

The first approach was to conduct a cross-sectional analysis of labour, material and plant costs and selected composite unit rates. Considering that a building is produced by a combination of material resources, processes and conditions, a suitable database of labour and material costs and unit rates for various building elements is required. The annual international construction cost and market survey by Turner & Townsend (2016) provided the necessary data for nearly 40 capital cities and regions around the world. The classification of these locations into lower-, middle- and high-income countries, based on their respective incomes per capita, facilitates analysis of these unit rates or comparisons between countries. Of the 27 countries examined, Qatar, Switzerland, Australia, Singapore, the US, Ireland, the Netherlands, Canada, Germany, the UK, the UAE, France, Hong Kong, South Korea, Oman, Chile, Poland and Russia were classified by the World Bank (2016) as high income (HI) in 2014; Brazil, Malaysia, Turkey, China and South Africa were classified as upper-middle income (UMI); and India and Kenya were classified as lower-middle income (LMI). The remaining two countries, Uganda and Rwanda, were classified as lower income (LI).

A second exercise was a longitudinal analysis based on labour costs, material costs and unit rates at a single location to test the applicability of this framework to changes in factor input costs. Wage rates, building material prices and unit rates for the city of Melbourne obtained from *Building Cost Guide*, published by Cordell Information Pty Ltd, provided the necessary data for this analysis. (A complete set of these quarterly publications dating back to Volume 1 published in July 1972 are available from the State Library of Victoria.)

In a third exercise the potential of the steel/concrete ratio for the preference of one material over another in 12 countries was carried out using data for tall buildings completed between 2000 and 2016 obtained from the Council on Tall Buildings and Urban Habitat (CTBUH 2016). Further inspection of the CTBUH time series data for Australia was carried out to elicit evidence of a switch from steel to concrete for these tall buildings.

Resource costs and cost ratios

The cost components analysed were limited to skilled labour cost (calculated as an average of skilled workers in three groups of trades), five basic material cost items (concrete, reinforcing bars, standard bricks, steel sections and softwood timber for framing) and unit rates for four key elements in the building trades (concrete in slabs, reinforcement in beams, formwork to soffit of slabs and structural steel sections). Labour costs were the all-inclusive cost to the employer, which included the basic hourly wage, allowances, taxes, annual leave cost and (where paid by the employer) worker's compensation and health insurance, pensions, and travel costs and fares. Material costs were for items delivered to site including all relevant taxes. Labour and material costs excluded overheads, margins, overtime and bonuses. The unit rates were the fully installed rates charged by the contractor to cover labour, materials, delivery, plant, overheads, margins and sales tax.

One of the challenges in analysing input costs internationally was that material, labour, plant and unit rates were all in their respective local currencies and required a suitable currency conversion to be applied to ensure comparability. An alternative was to divide an input cost such as labour or materials with another input cost, eliminating the effects of currency, exchange rates or PPPs. For example, the cost of one cubic metre of 30 MPa concrete could easily be expressed as a multiple of the hourly rate for skilled labour. Cost ratios for materials, labour and equipment were computed and analysed to compare the different proportion of these input factors for construction internationally.

The Turner & Townsend cost survey reported the hourly rates for skilled workers in local currencies and USD (based on the exchange rate prevailing in January 2016). It was apparent that the average hourly wage for skilled workers was highest in the high-income countries, and often very low in the middle- or low-income countries. A comparison of the average hourly rate in USD plotted against descending gross domestic product (GDP) per capita for the 27 countries is shown in Figure 8.1 (upper half). The data clearly shows that wages for skilled workers were high in high-income countries such as Switzerland, the US, Australia and the UK, and low in the lower-middle or lower-income countries. Breaking with this general trend were several high-income countries that exhibited lower wages relative to their GDP per capita. These were countries that engaged foreign construction workers in their domestic construction sector and paid them a wage rate commensurate with income levels in their country of origin rather than matching the wage rate of local workers. Most of the foreign workers in the construction sector in the Middle East and the city states of Singapore and Hong Kong originate from India, Bangladesh, Pakistan, Nepal, Indonesia, Sri Lanka and the Philippines and were paid wages that were significantly lower than the wages commanded by domestic workers (Al Awad 2010; CIDB 2014; Frost 2004; MoM 2016).

A similar international comparison of the cost of construction materials, for example, one cubic metre of concrete (shown in the bottom half of Figure 8.1),

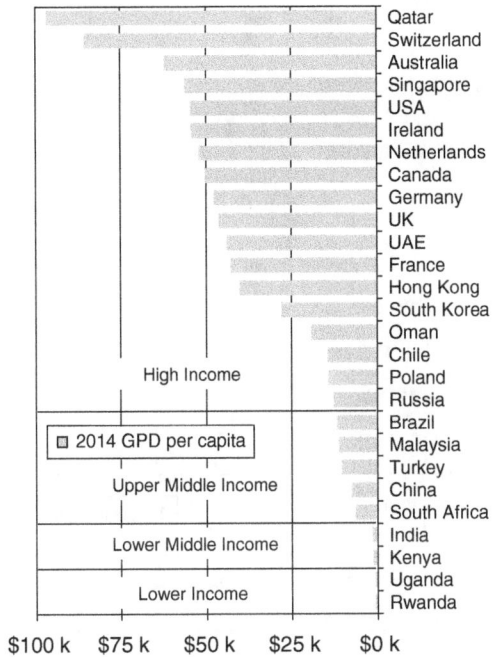

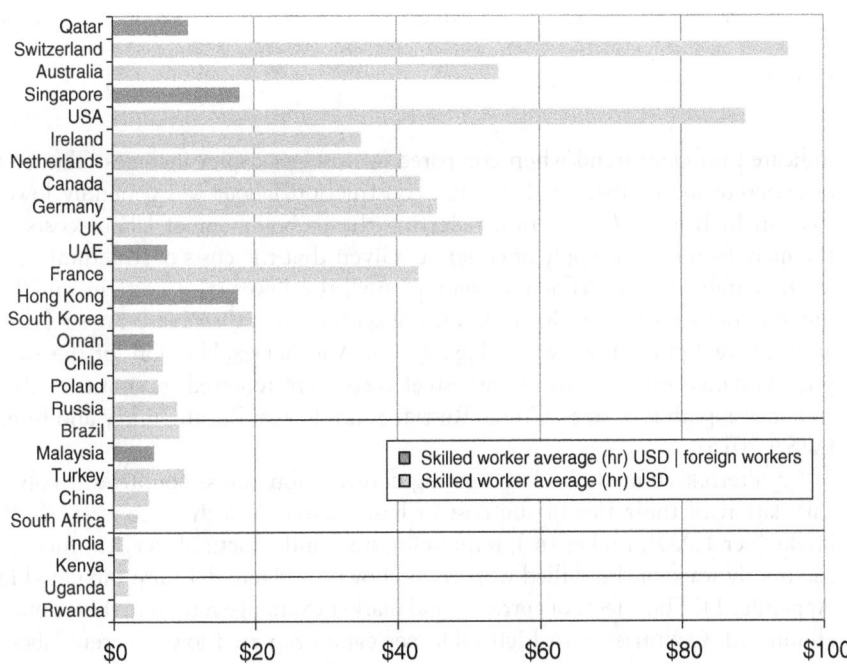

Figure 8.1 Average hourly wage for skilled workers and supply rate of concrete versus GDP per capita (all in USD) for high, upper-middle, lower-middle and lower-income countries (World Bank 2016; Turner & Townsend 2016)

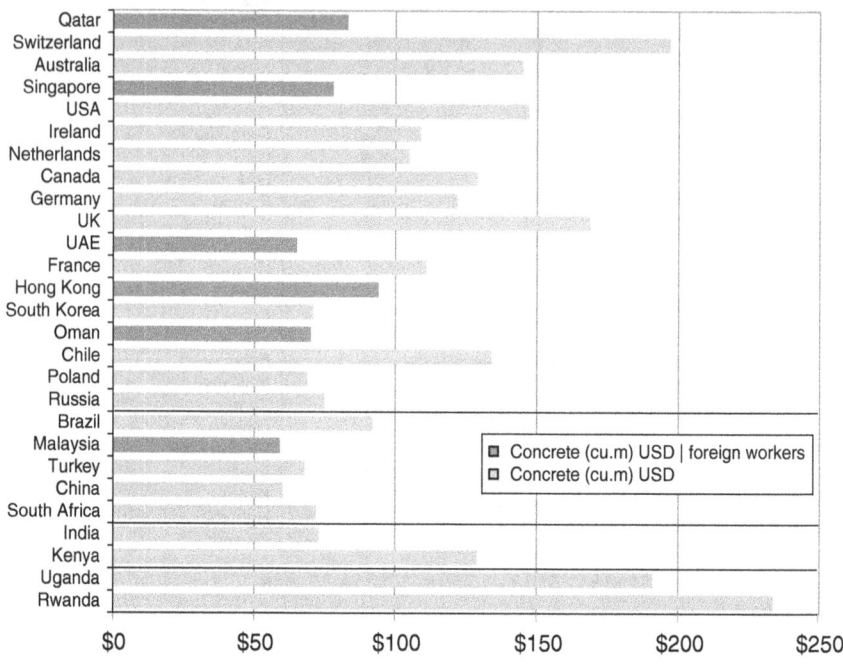

Figure 8.1 (Continued)

indicated no clear trend when compared against per capita incomes. The cost of concrete in countries with foreign construction labour was generally lower than in high-income countries, reflecting the embodiment of labour costs in the manufacture and supply of concrete. Given that the costs of structural steel in the database were defined as 'supply only', the lower steel costs in middle-income countries may be due to lower transportation or distribution costs (variation of steel prices is shown in Figure 8.2). Another explanation for the steel price variance was that the lowest steel costs were reported in countries that were net exporters of steel: China, Russia, South Korea, Brazil, Turkey and India (WSA 2015).

An alternative method of comparing construction across countries involves calculation of the ratios of the cost of basic materials such as concrete (m³), bricks (per 1,000), timber (m), reinforcing steel and structural steel (tonnes) to the hourly wage of the skilled workers as shown in Figure 8.3 (also tabulated in Appendix 1). The effects of currency and market exchange rate were then simply eliminated. Countries with high GDP per capita reported low concrete/labour ratios of approximately 2.0, whereas upper-middle-income countries reported an average ratio of 12.3. The cost of one cubic metre of concrete was equivalent to more than 60 hours of labour in lower-middle-income countries but increases rapidly to exceed 80 in lower-income countries. Supply of Ugandan concrete was the

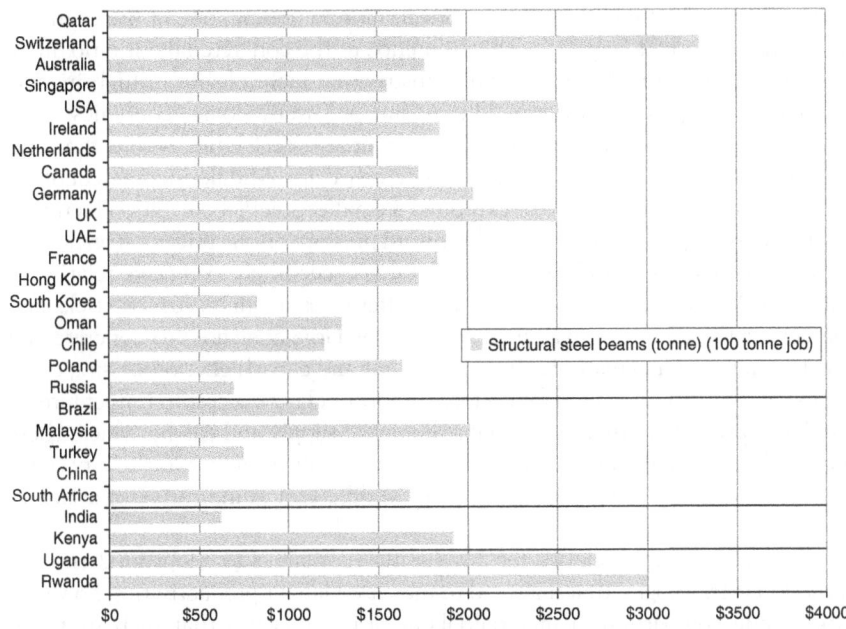

Figure 8.2 Supply rate (in USD) for structural steel (Turner & Townsend 2016)

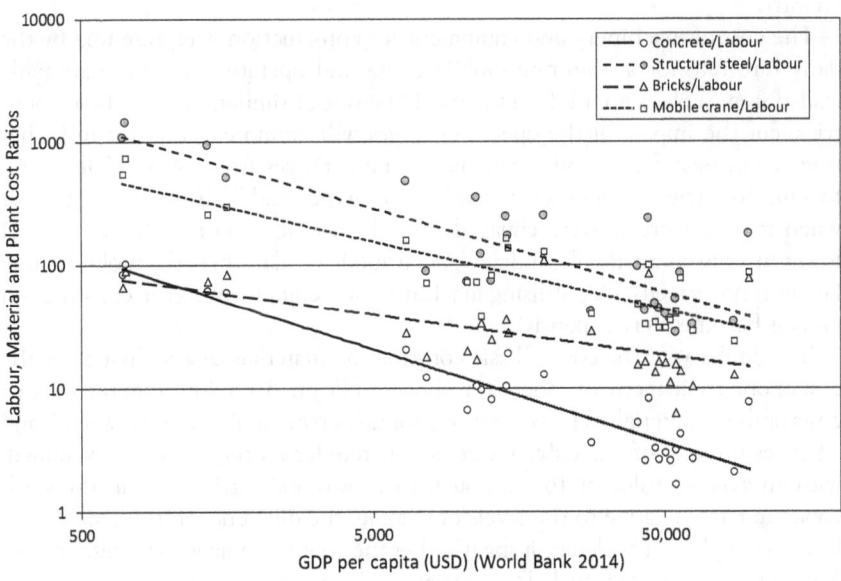

Figure 8.3 Ratio of materials and plant to labour (plotted on a log-log scale, with trend lines)

most expensive, at 102 times the hourly wage. High-income countries with many foreign workers exhibited lower wage levels more representative of the countries where these workers originate. The difference in the concrete/labour ratio drove the wider utilisation of labour-saving construction technologies in high-income economies compared to lower-middle-income or lower-income economies.

Similar patterns were established for structural steel ratios where the ratio approached a value of 30 for high-income countries. Middle-income countries that were net exporters of steel exhibited a smaller steel/labour ratio given that their excess supply of steel exerted downward pressure on the price of steel domestically. Figure 8.3 also indicates that steel was exceedingly expensive relative to labour in lower-income economies. The differences in the ratios for bricks and timber for the countries examined were not as large compared to concrete and steel, which require greater capital investment and carry higher processing costs. Bricks and timber, being reliant on local supply of clay or access to timber forests, are less costly where there is a reliable domestic supply and costlier where these materials must be imported. The higher relative cost of timber in the city states of Hong Kong and Singapore and in the Middle East is indicative of a lack of local supply.

There are multiple reasons for developing economies to adopt modern construction technologies that may lead to increased labour productivities; however, this simple comparison of material/labour ratios indicates that such strategies may not be cost-efficient if factors such as low wages, under-employment or capital shortages exist. It may be more advantageous to apply more labour-intensive processes and to utilise locally available or produced materials in a lower-income country.

The cost of machinery and equipment for construction is represented by the daily hire rate for a 50-tonne mobile crane and operator. An internationally traded item such as a mobile crane would be priced similarly across these countries, but the impact of the operator's wages will be more significant in higher wage countries. The mobile crane/labour ratio ranges from 24 to 40 for high-income countries, to between 80 and 170 in upper middle-income countries or when foreign workers were engaged in high-income countries. In the case of lower-income countries like India, Uganda and Rwanda, where the mobile crane/labour ratio exceeds 300, utilising machinery and equipment in the construction process becomes very expensive.

In a similar way, the cost of basic construction materials can be divided by the cost of one cubic metre of concrete as shown in Figure 8.4 to illustrate the relative costs of these materials. The cost for one tonne of structural steel is between 7 and 33 times the cost of one cubic metre of concrete for all the countries examined, with an average value of 16. Although there was a slight increase in the steel/concrete ratios relative to the levels of income, the difference in these ratios can be better explained by determining whether the country in question is either a net exporter or importer of steel. The average steel/concrete ratio was lower at 8.4 for net exporters of steel and markedly higher for net importers of steel. High steel output in China and India produced lower steel/concrete ratios of 7.31 and 8.50,

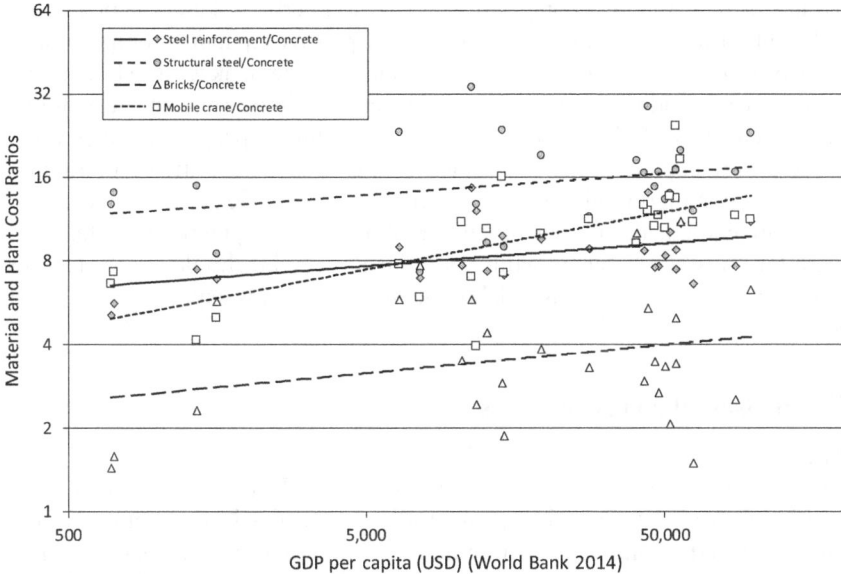

Figure 8.4 Ratio of materials and plant to material (plotted on a log-log scale, with trend lines)

respectively, indicating that construction in steel may be favourable in terms of cost compared to other countries where steel prices are significantly higher.

Bricks made by firing natural materials such as clay are relatively cheap in middle- and lower-income countries mainly as a result of the low labour costs involved in their production. It is clear that the brick/concrete ratio declined to a value less than 1.0 for lower-income countries, which suggests that wider use of bricks in lower-income countries is a cost-effective solution. In countries where bricks have to be imported, the relative cost can be extremely high, as shown in the case of Qatar, the UAE and Oman. Similarly, countries with a natural supply of native timber will report a lower timber/concrete ratio than countries that import timber.

Notwithstanding differences in the structure of the economy, climatic conditions, building regulations and standards and local preferences, the comparisons in the preceding sections may go some way towards explaining the choices made between building materials, and different proportions of labour and plant, in these countries. The selection of building materials based on the material cost ratio is clearly advantageous to lower-income countries where natural materials are abundant or the labour cost to manufacture or process these materials (e.g. cement and concrete) is inexpensive. Materials that are internationally traded (e.g. steel) would cost significantly more than locally produced materials in middle- or lower-income countries. The effect of lower-wage foreign workers in high-income countries drives the utilisation of materials with higher embodied labour (e.g. concrete compared to steel), and perhaps skews the preference to labour rather than to investment in more expensive equipment or machinery. Evidently, using more labour in lower-income countries makes sense from a cost perspective, in addition to the broader issue of providing or maximising employment, delivering skills training, and ultimately lifting people out of poverty. Based on the arguments above, builders in high-income countries will attempt to utilise construction methods that minimise the use of expensive labour, and substitute equipment and machinery for labour as far as practicable. The argument for utilising less labour while adopting advanced technologies or methods of construction that emphasise labour-saving practices is explored in the next section.

Comparison of composite rates

The composite rate for a 200 mm thick reinforced concrete suspended slab was computed from the unit rates obtained from the same Turner & Townsend (2016) cost survey, with labour and equipment allowances added for placement of the material into the building element plus a small allowance for wastage. Assuming a reinforcing steel quantity of 30 kg/m² of slab, the composite rates for a 200 mm thick slab were computed; the results are presented in Figure 8.5.

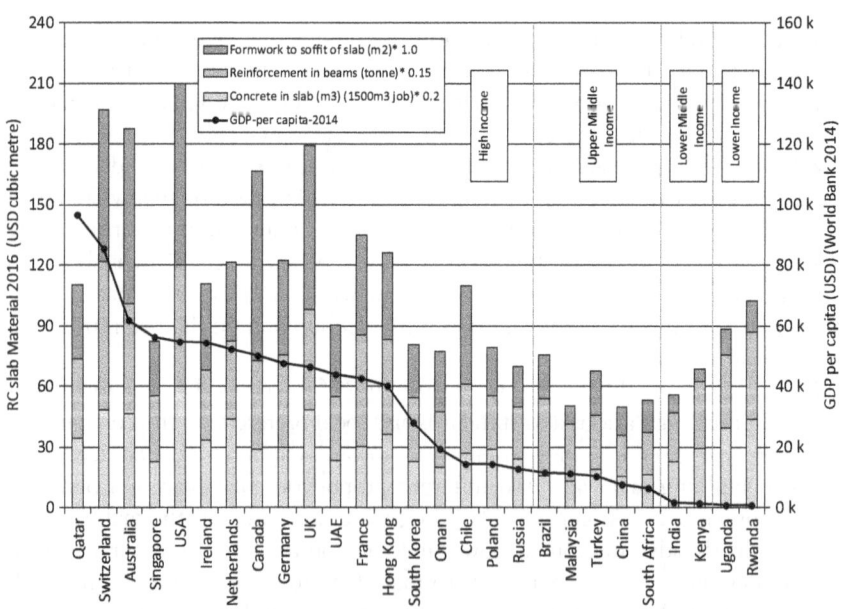

Figure 8.5 Composite rates (USD) for reinforced concrete slab (countries organised in descending GDP per capita)

It is apparent that the composite rates are higher in high-income countries due to higher material, labour and equipment costs. Formwork costs in high labour cost countries exceed 40% of the total composite rate in Canada, Australia, the US, the UK and Chile as a result of the extensive labour input into the installation and subsequent removal of formwork. This percentage reduces rapidly to less than 20% in lower-income countries, falling to a minimum of just 9% in Kenya. This reduction in the relative cost of timber formwork suggests that conventional in-situ placement of concrete with timber formwork remains the most cost-effective method of forming concrete in lower-middle-income and lower-income countries. Builders in high labour cost countries naturally seek less labour-intensive means of forming concrete, either with modular formwork systems, or table forms. They often select precast concrete or prestressed concrete alternatives to reduce labour, even though these systems may utilise more elaborate lifting equipment or expensive materials such as prestressing wires or cast-in sheet metal forms. The practice of engaging lower-wage foreign workers in high-income countries has the effect of reducing the overall composite rate for the conventional in-situ concrete slab, reducing the incentives for builders to adopt more capital-intensive alternatives.

Material choice for tall buildings

Data was obtained from the Council on Tall Buildings and Urban Habitat (CTBUH 2016) database of tall buildings to explore the link between the utilisation of structural steel or concrete for the main structure of these buildings. The analysis was conducted for buildings completed between 2000 and 2016 in 12 countries; results are presented in Table 8.1. Steel structures included both structures where the main elements were constructed from steel with a floor system of concrete on top of steel beams, and composite structures with a combination of steel and concrete components used together in the main structural elements. Examples include buildings which utilise steel columns with a floor system of concrete beams, steel structure with a concrete core, or concrete-filled steel tubes.

Countries with high structural steel/concrete ratios such as the UAE and Singapore reported low structural steel utilisation. Other high-income countries with mid-range ratios such as the United States and Germany reported structural steel utilisation of 21% and 20%, respectively. In comparison, Australia, Turkey, Russia and India with lower steel/concrete ratios were observed to prefer constructing in concrete. The stark difference in material use between UK and Canada, despite exhibiting approximately similar steel/concrete ratios, will require further investigation. Out of the group of steel exporting countries, South Korea and China registered high structural steel utilisation with 17% and 50% of the tall buildings being constructed of steel, respectively. Other steel exporting countries, such as India and Turkey, reported very low structural steel utilisation rates. These observations seem to support the utility of the steel/concrete ratio to explain the choice of steel or concrete, but other factors such as steel production capacity or a national culture of building may also have significant effects.

Table 8.1 Number and percentage of structural materials for completed tall buildings in various countries from 2000 until 2016 (CTBUH 2016)

Country	United Arab Emirates	Singapore	USA	Germany	UK	Canada	Australia	South Korea	Turkey	Russia	India	China
Structural steel/concrete ratio materials	28.8	20.0	17.1	16.7	14.8	13.3	12.2	11.6	11.0	9.3	8.5	7.3
Steel, composite	10	9	93	10	25	6	14	43	2	5	2	196
Concrete	326	98	342	37	28	242	222	204	75	442	96	186
Other	5	1	33	0	3	1	3	3	0	0	1	8
Total	341	108	468	47	56	249	239	250	77	447	99	390
Median height (m)	163	135	146	88	90	118	109	163	143	82	133	220
Percentage steel	3%	8%	20%	21%	45%	2%	6%	17%	3%	1%	2%	50%
Percentage concrete	96%	91%	73%	79%	50%	97%	93%	82%	97%	99%	97%	48%

Changes in technology over time

Labour wages, material costs and unit rates were collated for the city of Melbourne to examine how relative costs of labour and different materials may have influenced the choice of material or technology over a period. Data was obtained from a series of Australian building cost books (Cordell Building Information Services, various years 1972–2015). More recent editions of these cost guides contain a larger number of construction activities and unit rates for a broader range of building elements than did earlier editions. The purpose of this analysis was not to explain changes in technology for a wide range of building elements, but merely to show how these cost ratios might justify the selection of one method of construction over another. Sufficient information was available from these cost guides to examine five different concrete flooring systems, with the more recent post-tensioned concrete system first appearing in the 1995 edition. Data from early editions was converted to common units (e.g. converting concrete from cubic yards to cubic metres) for comparability.

Basic prices for labour, concrete, steel reinforcement, structural steel and mobile crane were obtained for the city of Melbourne from 1972 until 2015. Taking 1972 as the base year, these were converted to cost indices (see Figure 8.6) to illustrate the relative cost increase. The average hourly cost for a skilled worker increased approximately 6.4% annually over a period of more than 40 years. In comparison, the cost of one cubic metre of concrete only increased by 5.1%

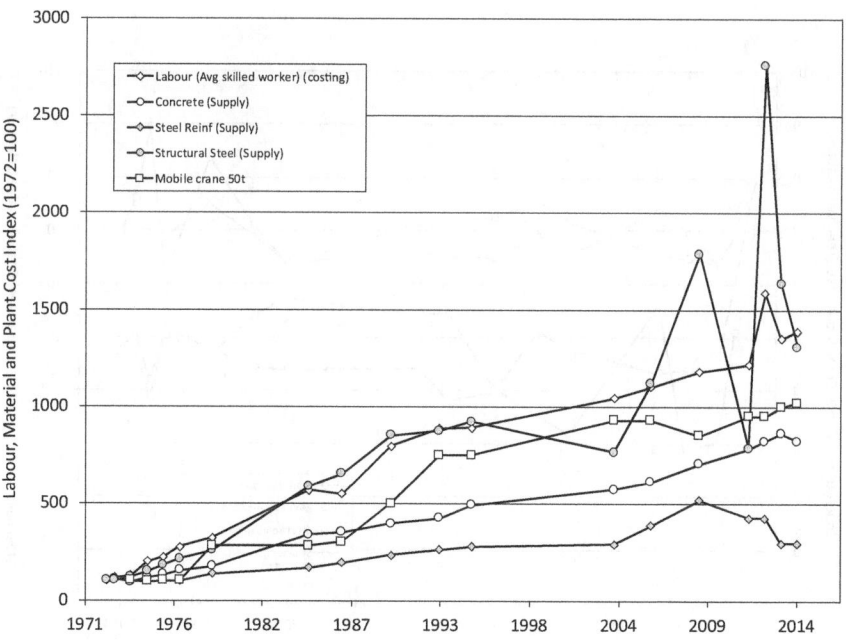

Figure 8.6 Basic labour, materials and plant cost indices for Melbourne (1972 = 100)

annually. The costs for steel reinforcement and structural steel, being commodities, exhibited a greater variance, especially between 2006 and 2015 following the price of steel traded in the global market. The cost of the hourly rental of a mobile crane increased 5.8% annually.

The ratios of basic materials and plant to labour using the methodology described earlier are plotted in Figure 8.7 to illustrate how these ratios change over this period of study. The concrete/labour ratio was observed to decline between 1972 and 1979 from 7.99 to 4.31 and remained generally at this level until 2015. In contrast, the ratios for steel reinforcement and structural steel were more stable at between 40 and 50 except for the period when the steel price was fluctuating wildly. The cost ratio for the mobile crane fluctuated between 3 and 5 with no discernible trend over time.

The availability of this longitudinal data provides a means to test the ability of these cost ratios to explain the choice of technology in two areas: (1) concrete slab construction and (2) structural framing in tall buildings. In the first comparison, the composite rates for five different reinforced concrete structural systems for a suspended slab were calculated. The only system described in the 1972 edition of the *Building Cost Guide* was a reinforced concrete suspended slab, 200 mm thick, concrete grade 32 placed by pump, steel reinforcement at a rate of 150 kg per m^3 of concrete, class 3 form-ply formwork and a steel trowelled finish. The reinforced concrete suspended slab with galvanised ribbed steel permanent formwork was included in the 1975 edition, with other systems such as post-tensioned prestressing and precast hollowcore added over the years. A plot of the relative costs of these

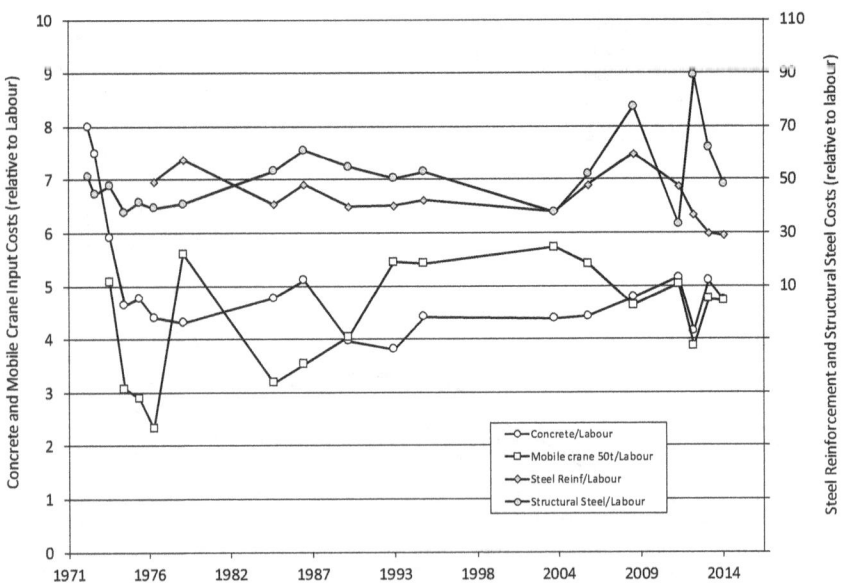

Figure 8.7 Ratios of materials and plant to labour for Melbourne from 1972 to 2015

systems is shown in Figure 8.8 together with the labour and concrete rates, and ratios of concrete/labour. With an initially lower labour cost, the concrete/labour ratio was at a peak at 7.99 in 1972 but this figure gradually fell to 5.11 in 1987. During this period, the more expensive system was the reinforced concrete slab with sheet metal formwork. As the result of an increase in the labour rates in 1990, the concrete/labour ratio fell to 3.96 leading to a sharp increase in the cost of the reinforced concrete slab with form-ply forms. Although there was a corresponding increase in the cost of the reinforced concrete slab with sheet metal formwork, the increase was not as large as for the more labour-intensive timber forms. From 1990 onwards, the rate for the sheet metal formwork remained lower than that for timber forms. The introduction of prestressing and precasting led to even more economical construction solutions, as costly labour-intensive processes such as the laying and

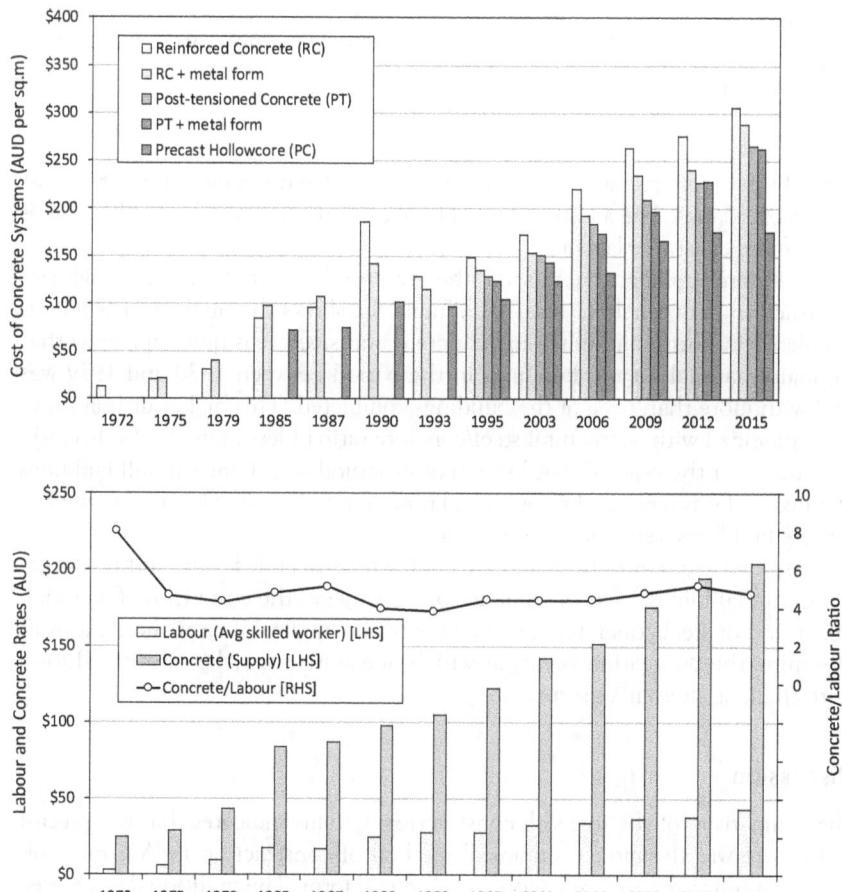

Figure 8.8 Increase in wages and concrete supply rate compared to five flooring systems in Melbourne (note that x-axis is not to scale)

Table 8.2 Number and percentage of structural materials for completed tall buildings in
       Australia from 1970 until 2016

Year	1970	1975	1980	1985	1990	1995	2000	2005	2010	2015
	1974	1979	1984	1989	1994	1999	2004	2009	2014	2016
Structural steel/ concrete ratio (average for five-year period)	6.8	8.8	–	11.5	13.5	12.0	8.6	13.9	13.3	10.2
Materials										
Steel, composite	14	28	8	7	12	5	1	5	5	3
Concrete	5	14	27	24	39	34	54	68	71	29
Other	1	1	0	0	0	1	0	0	0	0
Total	20	43	35	31	51	40	55	73	79	32
Percentage steel	70%	65%	23%	23%	24%	13%	2%	7%	6%	9%
Percentage concrete	25%	33%	77%	77%	76%	85%	98%	93%	90%	91%

Source: CTBUH.

tying of steel reinforcement were replaced by a post-tensioning system or precast
hollowcore planks. The additional cost of materials was easily offset by the savings
on the more expensive labour.

The second comparison was for the structural material used for high-rise
construction between 1972 and 2015. Table 8.2 shows the number of buildings
completed in Australia built using concrete and steel. It is quite apparent that
the main material for tall buildings in the period between 1970 and 1979 was
steel with more than 65% of the buildings completed. This higher utilisation of
steel coincided with a structural steel/concrete ratio of less than 10. As this ratio
increased over the years, the utilisation of structural steel frames in tall buildings
declined and was replaced by concrete. From year 2000 onwards, more than 90%
of these buildings were concrete structures.

These two examples of technology choice for suspended slabs and the structure of tall buildings, as a consequence of a change in the cost ratios of material/
labour and of steel/concrete, are evidence of the applicability of these ratios to
determine the bifurcation points at which one system may be substituted for a
more efficient alternative system.

Discussion

The comparison of the five slab construction systems indicates that the precast
hollowcore was the most economical method of construction in Australia followed by prestressed concrete with a sheet metal form. The traditional reinforced
concrete slab cast with timber forms was the most expensive, mainly due to the
highly labour-intensive nature of fixing and dismantling formwork combined
with high hourly wages, leading to a labour cost component that constitutes up to
50% of the total cost of the reinforced concrete slab (Yong 2010). It is therefore

not surprising that with the high cost of labour in Australia and other high-income countries, this construction system works out to be costlier compared to a precast solution.

Pricing the same project using the labour and material rates in middle-income Malaysia resulted in a lower proportion of labour costs: only 22% of the total cost of the reinforced concrete slab. This resulted in a situation where the cast in-situ system was the more economical solution compared to the precast system, which was 9% more costly. The higher manufacturing, transportation and erection costs for precast components more than outweighed the decrease in labour requirements for a precast solution in Malaysia. The total system cost for precast was 1.6 times higher than the in-situ reinforced concrete system, negating the obvious advantage of shorter construction schedules, increased productivity or improved quality.

No longitudinal data was available to allow an analysis that might determine the bifurcation point at which a precast system becomes more economical compared to a conventional cast in-situ concrete. Evidence from the Turner & Townsend (2012) data suggests that both these systems are at parity when the concrete/labour ratio is approximately 4. Builders in countries with low concrete/labour ratios (e.g. due to high wages) will seek to adopt labour-saving measures such as prefabrication, precasting, or invest in new technologies, whereas a high material/labour ratio suggests the adoption of more labour-intensive construction processes to reduce investment in complex equipment and machinery or reduce the utilisation of expensive construction materials.

In comparison to a cost performance study by Mills (2009) on medium-rise commercial buildings, the basic factor inputs and cost ratios proposed here have clearly demonstrated how labour and material costs affect the choice of construction systems. Using a standard model, Mills computed the costs of six structural framing systems in the five capital cities in Australia, and concluded that structural steel was by far the most expensive form of construction. This was attributed predominantly to the higher cost of steel plus the additional cost due to the requirement for fire protection to the steel sections. The study, coinciding with the high cost of steel in September 2008, indicated that despite the advantage of shorter construction schedules and the lighter total weight of a steel building, there was little incentive for using steel in construction. In fact, the Steel Reinforcement Institute of Australia (1994) reported that approximately 70% of Australia's 100 tallest buildings were constructed of reinforced concrete, lending further weight to the choice of reinforced concrete as the preferred method of construction. However, a detailed analysis of the same tall building database (CTBUH 2016) indicates that in 1970 to 1979, there were more buildings built with structural steel than reinforced concrete, when the ratio of the cost of structural steel to concrete was less than 10. The decline in the use of steel for tall buildings coincided with an increase of steel cost relative to concrete. Once the ratio exceeded 10 in 1985, the percentage of buildings built with steel dropped from 23% to a low of less than 10% beyond year 2000. This observation is corroborated by a per square metre cost of AUD 769 for a steel frame compared to AUD 499 for a reinforced concrete frame, based on costing in 2008 (Mills 2009).

These studies have shown how simple ratios can be utilised to justify the adoption of more prefabrication in a high-wage, high-income country like Australia while more labour-intensive construction processes are preferred in lower-wage, upper-middle-income Malaysia. Many other examples of these options can be found if comparisons are made between systems used in high-income and middle-income countries. Taking all basic construction materials together in this comparison indicates that at locations where labour is relatively cheap, it will certainly be worthwhile to adopt more labour-intensive construction processes and reduce reliance on expensive materials and/or plant and equipment. The higher wage costs in a high-income economy will evidently motivate builders to reduce their dependence on labour by adopting labour-saving options such as standardisation, prefabrication or precasting, even though these options may result in higher material or plant costs.

The representation of the factor inputs for construction as material/labour or material/material ratios has been inspired by the Cobb-Douglas production function, where output is estimated for various labour and capital inputs. The extent to which a firm has the ability to substitute between two different inputs at will, typically labour and capital, in order to produce the same level of output is known as an isoquant. In the construction industry this economic concept can be extended to two different materials, for example steel and concrete, which are substitutes. The findings clearly indicate that higher wages for construction workers lead to wider utilisation of labour-saving construction technologies (normally as a result of increased capital inputs), possibly leading to increased output and improved productivity. Conversely, countries that exhibit lower labour cost may still find it economical to build using more labour-intensive methods of construction. This finding is in line with the recommendations by the ILO in the 1980s (Tajgman and de Veen 1998) to adopt more labour-intensive methods of construction as policy for overcoming under-employment and achieving national development objectives. In addition, this research has now provided quantitative evidence of the most relevant and 'appropriate' construction technology for a set of wage levels and material costs. The decision on labour-intensive versus labour-saving construction technologies could be made on the basis of material/labour ratios while the selection of the appropriate construction material can be made on the basis of material/material ratios.

These cost comparisons provide a new perspective on international cost comparison efforts, and a deeper understanding of the ratios will enable construction economists to further refine their advice to clients on international construction costs and assist multi-lateral agencies to craft policy advice for developing countries that ensures a balanced development path for their construction industries.

Conclusion

A rational approach to the complex question of an 'appropriate' technology for construction has been proposed and demonstrated. The cost ratios presented were shown to be successful in determining the points at which various alternative technology options for building a suspended reinforced concrete slab are chosen in different countries/economies. Similar ratios can explain the selection of, for example, concrete or steel as the main structural material for tall buildings.

Cost ratios were derived from basic factor inputs such as labour costs (i.e. wages of skilled workers) and material costs (i.e. supply costs of common construction materials such as concrete, steel, bricks and timber), together with rental costs for plant and equipment (i.e. mobile cranes). Data from 27 countries was examined, ranging from high-income, upper-middle-income, lower-middle-income and low-income countries as defined by the World Bank. The effect of employment of foreign workers in a number of high-income and upper-middle-income countries is to shift the wages of these workers to a level closer to that of their country of origin than that of the host countries. The consequence of these lower wages is a set of composite trade rates that are more akin to middle-income countries and a preference for labour over capital in their respective production functions.

This cost ratio analysis may help provide construction professionals and researchers with a practical quantitative tool to evaluate 'appropriate' options for construction industry development and economic growth, especially for rapidly developing economies facing constraints of labour, capital or resources.

References and further reading

Al Awad, M. (2010) *The Cost of Foreign Labor in the United Arab Emirates*, Institute for Social and Economic Research, Zayed University, Working Paper No. 3, July 2010, UAE.

Cobb, C. W. and Douglas, P. H. (1928) A theory of production. *The American Economic Review*, **18** (1), 139–165.

Cordell Building Information Services. (various years) *Cordell Commercial and Industrial Building Cost Guide – Victoria*. Various editions from July 1972 to January 2015.

CIDB (2014) *Wage Rates of Construction Workers*, January 2014 (Daily). Construction Industry Development Board, Malaysia. http://myn3c.cidb.gov.my/cidb_n3c/progress/index.php.

CTBUH (2016) *The Skyscraper Center – The Global Tall Building Database of the CTBUH* (Council on Tall Buildings and Urban Habitat). http://skyscrapercenter.com/, 9 September 2016.

Douglas, P. H. (1967) Comments on the Cobb-Douglas production function. In: Brown, M. (ed.) *The Theory and Empirical Analysis of Production* (New York: National Bureau of Economic Research).

Frost, S. (2004) Building Hong Kong: Nepalese labour in the construction sector. *Journal of Contemporary Asia*, **34** (3), 364–376.

Ganesan, S. (1979) Growth of housing and construction sectors – Key to employment creation. *Progress in Planning*, **12**, 1–79.

Ganesan, S. (1994) Employment maximization in construction in developing countries. *Construction Management and Economics*, **12** (4), 323–335.

Ganesan, S. (2000) *Employment, Technology and Construction Development: With Case Studies in Asia and China* (Farnham, UK: Ashgate Publishing).

Gaude, J. and Watzlawick, H. (1992) Employment creation and poverty alleviation through labour-intensive public works in least developed countries. *International Labour Review*, **131** (1), 3–18.

Hillebrandt, P. M. (1974) *Economic Theory and the Construction Industry* (New York: Springer).

Hillebrandt, P. (1999) Choice of technologies and inputs for construction in developing countries. In: *Construction Industry Development in the New Millennium, Proceedings of the Second International Conference on Construction Industry Development*, National University of Singapore, October.

ILO (1998) *Employment Intensive Infrastructure Programmes: Labour Policies and Practices* (Geneva: International Labour Organization).

ILO (2001) *The Construction Industry in the Twenty-First Century: Its Image, Employment Prospects and Skill Requirements* (Geneva: International Labour Organization).

ILO (2006) *Independent Evaluation of ILO's Strategy for Employment Creation Through Employment-Intensive Investment Approaches (EIIS)* (Geneva: International Labour Organization).

Mills, A. (2009) Cost performance of multi-rise structures in Australia. *Building Economist*, September, 10 (Canberra: AIQS).

MoM (2016) *Singapore Yearbook of Manpower Statistics 2016*. Manpower Research and Statistics Department, Ministry of Manpower, Republic of Singapore.

Moavenzadeh, F. and Rossow, J. A. K. (1975) *The Construction Industry in Developing Countries*. Technology Adaptation Program, Massachusetts Institute of Technology.

Robinson, J. (1953) The production function and the theory of capital. *The Review of Economic Studies*, **21** (2), 81–106.

Steel Reinforcement Institute of Australia (1994) Australia's 100 Tallest Buildings. *Reinforced Concrete Digest*, **D3**, September.

Sultan, B. and Kajewski, S. (2006) Requirements for economic sustainability in the Yemen construction industry. In: Serpell, A. (ed.) *Proceedings International Symposium on Construction in Developing Economies: New Issues and Challenges*, Santiago, Chile.

Tajgman, D. and de Veen, J. (1998) *Employment-Intensive Infrastructure Programmes: Labour Policies and Practices* (Geneva: International Labour Office).

Tatum, C. B. (1988) Classification system for construction technology. *Journal of Construction Engineering and Management*, **114** (3), 344–363

Turner & Townsend (2012) *International Construction Cost Survey 2012* (Brisbane: Turner &Townsend).

Turner & Townsend (2016) *International Construction Market Survey 2016* (Brisbane: Turner &Townsend).

Warszawski, A. (1999) *Industrialized and Automated Building Systems* (London: E. & F.N. Spon).

World Bank (2016) *GDP Per Capita (current US$)*. 14 June 2016. http://data.worldbank.org/indicator/NY.GDP.PCAP.CD.

WSA (2015) *World Steel in Figures* (Brussels: World Steel Association).

Yong, T. N. (2010) *Feasibility of Precast Concrete Construction System in Malaysia: A Comparative Study Between Australia and Malaysia*. Research project report, The University of Melbourne, October.

Chapter 8 Appendix

Table A1 Labour and material data (in USD), derived ratios and composite rate for suspended slab

	Qatar	Switzerland	Australia	Singapore	USA	Ireland	Netherlands	Canada	Germany	UK	United Arab Emirates	France	Hong Kong	South Korea
Labour costs														
Skilled worker (average of three)	10.67	95.00	54.33	18.00	89.00	35.00	40.67	43.33	45.67	52.00	7.67	43.00	17.67	19.67
Material costs														
Concrete 30 MPa (m³)	83	197	145	78	147	109	105	129	122	169	65	111	94	71
Reinforcement bar 16 mm (t)	922	1507	956	846	1100	957	1072	1079	931	1279	926	967	866	630
Concrete block per 1,000	1163	5370	2464	504	1300	652	2914	1115	3317	1338	954	2935	504	841
Standard brick per 1,000	842	1231	569	252	561	543	450	705	761	706	681	707	281	575
Structural steel beams (t)	1916	3295	1761	1551	2512	1848	1481	1727	2036	2500	1880	1837	1728	825
Softwood timber 100 × 50 mm (m)	5	5	2	9	5	5	2	4	3	5	4	3	9	2
Derived ratios														
Concrete/labour	7.68	2.07	2.67	4.26	1.65	3.13	2.60	2.98	2.65	3.25	8.28	2.59	5.27	3.63
Steel reinforcement/labour	85.30	15.86	17.59	46.42	12.36	27.50	26.41	24.86	20.24	24.62	117.24	22.63	48.76	32.06
Structural steel/labour	177.33	34.68	32.40	85.11	28.22	53.13	36.48	39.78	44.24	48.11	237.93	42.97	97.26	41.96
Bricks/labour	77.92	12.96	10.47	13.82	6.30	15.63	11.09	16.24	16.54	13.58	86.21	16.53	15.81	29.23
100 × timber/labour	48.31	5.26	4.00	47.37	5.62	15.63	5.36	9.94	7.09	11.32	44.83	7.63	53.03	12.00
Mobile crane/labour	86.44	24.08	29.33	73.95	40.45	42.19	35.41	31.16	30.73	34.53	100.00	33.05	48.63	40.71
Steel reinforcement/concrete	11.11	7.65	6.60	10.89	7.48	8.80	10.16	8.33	7.65	7.57	14.17	8.73	9.25	8.83
Structural steel/concrete	23.10	16.73	12.15	19.96	17.09	17.00	14.04	13.33	16.72	14.78	28.75	16.57	18.44	11.56
Bricks/concrete	10.15	6.25	3.93	3.24	3.82	5.00	4.27	5.44	6.25	4.17	10.42	6.37	3.00	8.05
100 × timber/concrete	6.29	2.54	1.50	11.11	3.40	5.00	2.06	3.33	2.68	3.48	5.42	2.94	10.06	3.31
Mobile crane/concrete	11.26	11.61	11.00	13.52	24.49	13.50	13.63	10.44	11.62	10.61	12.08	12.75	9.22	11.22
Composite rates														
RC slab 200 mm	110.34	196.91	187.76	82.59	210.00	110.83	121.45	166.51	122.34	179.22	90.09	134.84	126.30	80.42

	Oman	Chile	Poland	Russia	Brazil	Malaysia	Turkey	China	South Africa	India	Kenya	Uganda	Rwanda
Labour costs													
Skilled worker (average of three)	5.67	7.00	6.33	9.00	9.33	5.67	10.00	5.00	3.33	1.20	2.07	2.00	2.80
Material costs													
Concrete 30 MPa (m³)	70	134	69	75	92	59	68	60	72	73	129	191	234
Reinforcement bar 16 mm (t)	650	952	680	552	1111	876	522	414	650	504	964	1077	1192
Concrete block per 1,000	550	1457	1083	928	609	806	831	629	274	628	964	1292	1430
Standard brick per 1,000	580	199	239	235	322	97	203	89	99	102	154	170	188
Structural steel beams (t)	1300	1200	1637	698	1171	2004	746	437	1675	620	1926	2712	3001
Softwood timber 100 × 50 mm (m)	3	3	2	3	2	3	2	5	4	4	3	3	3
Derived ratios													
Concrete/labour	13.00	19.16	10.44	8.03	9.65	10.46	6.67	12.19	20.63	60.00	62.66	101.56	83.82
Steel reinforcement/labour	125.00	136.51	102.53	59.03	117.13	154.05	51.33	84.38	185.67	414.00	466.61	571.28	426.98
Structural steel/labour	250.00	172.07	246.84	74.65	123.39	352.54	73.33	89.06	478.53	510.00	932.75	1438.44	1075.11
Bricks/labour	111.00	28.50	36.08	25.08	33.91	17.11	20.00	18.13	28.34	84.00	74.53	90.00	67.27
100 × timber/labour	50.00	36.00	30.38	35.56	23.48	60.81	23.33	93.75	119.63	342.00	145.25	160.94	120.84
Mobile crane/labour	130.00	137.99	167.09	83.33	38.14	72.97	73.33	71.88	160.12	300.00	259.18	738.36	551.85
Steel reinforcement/concrete	9.62	7.13	9.82	7.35	12.14	14.73	7.70	6.92	9.00	6.90	7.45	5.62	5.09
Structural steel/concrete	19.23	8.98	23.64	9.30	12.78	33.71	11.00	7.31	23.19	8.50	14.89	14.16	12.83
Bricks/concrete	8.54	1.49	3.45	3.12	3.51	1.64	3.00	1.49	1.37	1.40	1.19	0.89	0.80
100 × timber/concrete	3.85	1.88	2.91	4.43	2.43	5.81	3.50	7.69	5.80	5.70	2.32	1.58	1.44
Mobile crane/concrete	10.00	7.20	16.00	10.38	3.95	6.98	11.00	5.90	7.76	5.00	4.14	7.27	6.58
Composite rates													
RC slab 200 mm	77.30	109.75	79.26	69.59	75.21	50.34	67.43	49.65	53.16	55.83	68.34	88.60	102.18

Editorial comment

In the following chapter the author describes and discusses the most recent full round of the International Comparison Program (ICP 2011). Having been part of the team that developed the current ICP methodology for producing construction industry-specific purchasing power parities (PPPs), the author, Jim Meikle, provides expert commentary based on first-hand experience.

The processes described demonstrate the continuing evolution of the ICP methodology as efforts continue to address the many complexities associated with the production of construction PPPs across a very diverse set of around 200 economies. With each round of the ICP, further analysis of the difficulties is possible as more data becomes available and further experience of data collection and validation accrues. There is increased understanding of how to frame and populate survey instruments, of precisely what data is needed and, perhaps more importantly from a purely practical perspective, what data can be collected and validated that will be at least reasonably consistent and reliable.

The ICP remains a work in progress, but the 2011 round arguably represents the most refined exercise for construction to date. As is often noted, however, there are no perfect or correct PPPs against which the latest results can be compared so it is not possible to know for certain if the results are truly an improvement on those from previous rounds. No doubt there is still plenty of scope for further analysis of methods, data and overall results, and such analysis may, in time, lead to a generally accepted approach that provides robust and dependable PPPs for construction.

9 A review of the 2011 construction survey and results from the World Bank International Comparison Program

Jim Meikle

Introduction

Making credible international price level comparisons is difficult. Commercial exchange rates do not necessarily reflect real differences in purchasing power between countries, and a single currency convertor does not accurately represent price level differences across different components of an economy. The World Bank, through its International Comparison Program (ICP), is responsible for the production of purchasing power parities (PPPs) for both national economies (gross domestic product, GDP) and for sub-components of GDP for around 200 countries (see World Bank 2018a). PPPs are alternatives to market exchange rates and are intended to reflect price level differences across countries more accurately. One of the sub-components of GDP in the ICP is construction, part of gross fixed capital formation (GFCF) or investment.

The main components of GFCF are machinery and equipment, and construction. In terms of content and price levels these two are very different: machinery and equipment items are generally internationally traded and as a result are likely to have PPPs that are broadly similar to commercial exchange rates; the bulk of construction, on the other hand, is an essentially local activity and is likely to have PPPs that are markedly different to exchange rates or machinery and equipment PPPs. In poorer countries, construction price levels and, therefore, construction volumes are likely to be understated using exchange rates, while in richer countries the opposite is often the case.

The history of the development of PPP theory and its application to construction is described, as is the evolution of the calculation methods for construction PPPs, in Best and Meikle (2015: see chapters 2–4).

Construction is described by the World Bank and other international agencies as 'comparison resistant'. According to the Organisation for Economic Co-operation and Development (OECD) *Glossary of Statistical Terms*, this is a term used to describe goods and services whose complexity, variation and country specificity make it difficult for them to be priced reliably across countries (OECD 2007). This is commented on throughout the chapter, but it is important to emphasise from the start that the calculation of PPPs for construction is problematic, much more so than it is for many other economic activities.

To date the ICP has undertaken regular but relatively infrequent international price surveys; the last two were in 2005 and 2011. In 2016, the United Nations Statistical Commission agreed that the ICP should become a permanent element of the global statistical system and should be conducted more frequently (World Bank 2016).

The detailed results of the 2011 survey were published in October 2015 and this chapter is largely based on the 2011 survey and its results (World Bank 2015a). The chapter describes and discusses the methods adopted in, and the results from, the 2011 ICP construction survey; it is in four sections including this introduction. The second section, within the limits of the author's information and understanding, summarises the main elements of the approach adopted by the ICP and describes how and why work on it evolved as it did (the author acted as a consultant to the ICP on the construction survey from 2009 to 2013). The third section presents and comments on selected results from the ICP 2011 survey: construction PPPs, construction price level indices (PLIs) and construction expenditures. The final section draws conclusions from the survey and its results and makes suggestions for the future conduct and analysis of international construction price surveys.

The approach to the ICP 2011 survey

PPPs are spatial price indices – they measure price differences across locations – and like other price indices, their calculation calls for an appropriate list of items, prices for these items and weights that represent the contribution of each item or group of items to the activity being measured. Three key decisions were made at an early stage of development of the ICP 2011 construction survey that helped shape the approach adopted subsequently:

- that the work would be based on construction resources (primarily materials and products, labour and hire of construction equipment), i.e. construction inputs rather than outputs such as construction projects;
- that input prices paid by contractors to suppliers for construction resources would be collected, rather than output prices (prices paid by purchasers for completed construction work); and
- that the aim would be to produce PPPs for different types of construction work and for all construction work directly rather than via construction projects.

The rationales for these decisions emerge in the text that follows.

The broad approach adopted for the ICP 2011 construction survey and how that evolved over the period 2009–2013 is described below. It comments on the selection of items, the collection of prices and the choice of weights (for reviews of alternative and previous ICP methods for construction surveys, see chapter 4 in Best and Meikle 2015). This chapter does not cover the period of final production of PPPs.

Selection of items

The items included in the construction price survey were selected as being representative of most types of construction work, relatively straightforward to describe and in common use across most countries. A total of 50 items was selected – enough to give a reasonable spread and not so many as to make collecting prices too burdensome for respondents – 38 materials and products, seven types of labour and five types of construction equipment (machinery such as excavators, cranes, etc. used during construction works). Best undertook a survey of published price data that informed the selection of items (for a summary of that work see the appendix to chapter 4 in Best and Meikle 2015). Some effort was also made to link items in the ICP 2005 and 2011 construction price surveys. The selected items and their brief descriptions are listed in the Appendix to this chapter. Additional notes and images for materials and products and equipment were prepared by the World Bank Global Office and provided to survey respondents.

The survey form permitted respondents to price alternatives where items specified in the survey documents were not commonly available but local equivalents were, for example, common sand and cement bricks could replace common clay bricks; commonly used hardwoods could replace softwood and so on. Preferred units were provided for all items – for example, cubic metres for sand and aggregates, square metres for plywood, days for bricklayers – but provision was also made for respondents to insert other units in common use locally. This involved those checking and analysing survey responses to convert prices for items based on local units to prices for standard units. Experience from the 2011 survey indicates that the survey instrument, including the selection of items, item descriptions and supporting notes, could all be improved but that the general approach was broadly satisfactory. The Appendix also indicates, with coefficients of variation (CoVs), which items were more variable in their pricing than others; the CoVs and their significance are discussed in more detail below.

Collection of prices

ICP-type exercises call for the comparison of prices of comparable products or services in each country; ideally these items should be as close to identical as possible. Most consumption price data is based on multiple observations of retail prices paid directly by end users for more or less identical products or services – packets of cornflakes, tubes of toothpaste or haircuts, for example. With few exceptions, this is not possible for construction; comparable, never mind identical, construction projects are difficult, if not impossible, to find.

It is difficult to observe any construction prices, but reliable output prices are particularly problematic as they are only available for completed projects and these will always incorporate context, locational, temporal, site and project-dependent factors that can significantly impact on price levels and comparability. Examples include climatic and seismic conditions, market factors, site access and ground conditions. In addition, projects will always be designed and built to comply with local standards, regulations and practices, and prices will reflect that. And finally, comparable units of measurement do not exist for many construction types. The majority of construction projects are more or less one-offs and, while they may well be representative of their country or location of origin, they are not strictly comparable across countries.

Reliable input prices are also difficult to collect but less so than output prices. The decision was made, therefore, to concentrate on input prices and, if possible, adjust these to approximate output prices. A major advantage is that prices for standard units of purchase – cubic metres for concrete, square metres for plywood, days for the hire of labour, for example – are available. Published input prices such as official labour rates and material price lists, however, are indicative only and are not appropriate for many types of construction work. Large projects can attract substantial quantity discounts and large contractors can obtain discounts regardless of the size of any particular project; smaller projects and smaller contractors will often pay significantly higher prices for construction resources. Collecting representative input prices, therefore, calls for care and experience.

Provision was made in the survey form for adjustments for some or all of the regional variations (where other than national average prices were provided), contractors' mark-ups (for site and head office overheads and profit) and professional fees. Data for mark-ups and professional fees was collected but not used due to data gaps and concerns about data quality. The approach followed for filling gaps in price data is explained in detail in chapter 19 of the ICP *Operational Guidelines* (World Bank 2015d).

National construction experts were selected as the primary sources of price data (i.e. government employees, industry researchers or private consultants). This generally meant that only single price observations were reported in each country. However, national experts, if chosen sensibly, bring broad experience of different types of construction in different locations and circumstances. All other construction PPP methodologies use national construction experts, although usually for output and/or project prices. External experts take time to identify, appoint and brief, and require payment. The ICP 2011 timetable was very condensed and survey preparation and validation may have benefited from additional time.

The ICP calls for annual average and national average prices, although mid-year prices were accepted instead of annual averages. National averages (i.e. an average of prices charged throughout a country) are asked for in the ICP survey, although sometimes prices were provided for specific locations – usually capital or main cities. Where this was done, respondents were also asked for a factor to convert prices submitted to national average prices. More time spent training respondents should improve response rates and the quality of responses.

The survey form was designed so that different 'baskets' of materials could be compiled for each of three types of construction work – residential and non-residential building and civil engineering. These baskets are termed by the World Bank as basic headings, components of the economy for which PPPs are calculated, however only one construction PPP is published, for 'All construction', an expenditure weighted aggregate of the three basic heading construction PPPs. For example, cement and steel are commonly used in all types of construction, roof tiles are used in building work and not in civil engineering. Respondents were asked to select items that were considered locally 'important' for each type of work/basic heading. Importance was defined as items that were readily available and commonly used. The Global Office established default selections for those countries that could or did not (see Appendix).

'All-in' prices were also collected for different types of construction work, for example, per square metre for buildings or per metre run for drains. Although this method of estimating project prices is commonly used in many countries, the results are not very reliable as there are different inclusions and exclusions, different rules of measurement, and the prices provided are for projects representative of each country which are not necessarily comparable across countries. Not all respondents completed this section of the survey and the data collected was not used.

Checking and validation of the construction survey data was generally undertaken by large groups of national statisticians that reviewed a range of survey results across all elements of GDP. This author's personal experience from the Eurostat construction price surveys suggests that smaller groups involving construction experts are useful to help resolve misunderstandings and arrive at acceptable price data.

It is difficult to independently and reliably assess the extent to which item prices or price levels vary from 'correct' values, not least because objectively correct values are not known. But variability in construction prices can be very high. A study for the World Bank indicated that, within a country, project estimates can vary by ±10%, trade estimates within a project can vary by up to 25% and individual items can vary by at least 50% (Sinclair *et al.* 2002). Variability across countries can be much higher. An exercise by Davis Langdon for Eurostat indicated that, across the members of the European Union in 2009, project prices (normalised by PPPs) varied by almost 100%, work group (trade or element) prices varied by more than 150% and individual items varied by factors of 6 to 20 (Davis Langdon n.d.). Submitted prices, therefore, should not be rejected or amended merely because they are subjectively considered to be 'too high' or 'too low'. Apparent outliers should be thoroughly checked with respondents to ensure that they relate to the particular item, that the items are in common use and that the correct units have been priced. Wherever possible, triangulation should be used to cross-check data and results from different sources and methods.

An analysis was undertaken of the variability, indicated by CoVs, of individual price levels from a group of around 100 countries (Thomas 2013). A similar approach was used by Best (2008) in analysing input costs from six locations. The CoVs for all resources are included in the Appendix and commented on below.

The CoV of each item measures how closely the price level of that item is to all the national price levels for that group of items. An item with a low CoV is a better proxy for national price levels than one with a high CoV. Item CoVs are calculated from item price levels (after conversion to a common currency) divided by overall country price levels for each group of items. Separate CoVs are calculated for materials and products, and labour and equipment items. Table 9.1 presents the materials and products with the highest CoVs and those with the lowest; it also shows low, high and average CoVs for labour and equipment items.

The materials and products overall indicate the greatest variation compared with labour and equipment items, even those with the lowest CoV. The labour items, with the exception of the machine operator, indicate the lowest CoVs and the equipment items are in between. The relatively low CoVs for both the labour and equipment items are not because their prices are more consistent across countries (they are not), but because they tend to be more consistent within countries.

The items with higher CoVs for materials and products are a mixture of complex manufactured and internationally traded items and electricity, influenced by exchange rates, transport costs and government policies; the lower CoV materials and products are commodity items, either locally produced or internationally traded. The lower CoV items are also generally those that are easier to specify, while at least some of the higher CoV items are more difficult to specify and/or more likely to be produced to local requirements.

It is difficult to conclude too much from this analysis other than that the simpler and easier to describe an item is, the more likely that its price level will be relatively consistent with other price levels in that item group in that country; and the more complex and country-specific an item is, the more likely it is to have different price levels. But the incidence of different price levels for construction items within a country is not unusual. Analysis like this is undertaken on Eurostat results as part of the validation exercise and can help identify possible outliers, but outliers are not necessarily wrong – they just require thorough checking.

The author's experience of a limited number of ICP 2011 validation meetings and reviews suggests that mechanical and electrical items (pumps, fans,

Table 9.1 Coefficients of variation for selected items (based on Thomas 2013)

Materials and products				Labour and equipment items	
Items with highest CoVs	CoV	Items with lowest CoVs	CoV	Selected items	CoV
Sheet glass	2.80	Ready mix concrete	0.64	Carpenter	0.11
Electricity	2.21	Structural steel	0.63	Electrician	0.19
Wash hand basin	1.88	Aggregate	0.59	Machine operator	0.36
Electric fan	1.60	Sand	0.56	*Average labour CoV*	0.19
Cast iron pipe	1.58	High yield reinforcement	0.54	Tandem vibrating roller	0.26
AC equipment	1.49	Mild steel reinforcement	0.52	Skid steer loader	0.31
Electric pump	1.41	Portland cement	0.52	Tracked tractor	0.35
Average materials CoV	1.07	Precast concrete slabs	0.51	*Average equipment CoV*	0.28

air-conditioning equipment and the like) and roofing materials (tiles and sheet) are problematic. Their specifications often tend to be country-specific and may require some adjustment to make them comparable. If possible, item descriptions should be improved, and in the most extreme cases the items should probably be omitted. There is also doubt whether petrol and diesel fuel and electricity should be included; they are probably not that significant as construction resources and their prices can be heavily influenced by national taxes or subsidies. Interestingly, some, but not all, of these items also have higher CoVs.

Choice of weights

During the preparations for the ICP 2011 survey, a number of methods were considered for weighting resources to represent different types of construction work (basic headings). The main ones considered (and rejected) were weights based on the mix of inputs in the construction column or columns of national input-output or supply and use tables and weights derived from model projects.

Input-output / supply and use table-based weights were rejected because not all countries had these kinds of presentations of national accounts and many of them were not considered sufficiently consistent in form and content, or sufficiently reliable or up to date. In addition, many national tables only have a single column for construction, although some have multiple columns (for residential buildings, civil engineering work, etc.). Tables with single columns for construction work can only be used to produce 'all construction PPPs'. This problem is reducing and will almost certainly continue to reduce over time. A recent African Development Bank (AfDB) survey, reported in the *African Statistical Journal*, indicated that 29 African countries have compiled at least one table since 2000 and 14 countries now compile them every year (AfDB 2015). The Asian Development Bank (ADB) has been assisting member countries in the production of tables for some time and 17 Asian countries now compile them (ADB 2018). The OECD publishes standardised input-output tables on an annual basis for 61 OECD and non-OECD countries (OECD 2018). The input-output based approach deserves further study, at least as a check on PPPs at the 'all construction' level.

Project-based weights were rejected because it was felt that coming up with a set of projects that would reasonably represent construction work and provide acceptable comparisons across the range of countries in the ICP was too difficult. And, in any case, 'projects' in these types of exercises are typically newbuild projects; it is extremely difficult to identify, describe and price refurbishment or conversion projects, and these can constitute a significant proportion of construction expenditure in many countries. There is also the problem of aggregating projects to types of work.

The method finally adopted for combining individual groups of resources (materials, labour and equipment) was, broadly, to calculate price relatives (effectively PPPs) for each resource item and then aggregate individual groups of resource items using geometric means. Aggregate resource PPPs were then combined into basic heading PPPs using the estimated shares (resource mixes) that each resource represents in each basic heading's output. Details of calculation methods are set out in the ICP *Operational Guidelines* (World Bank 2015d).

Resource mixes were generally provided by national experts, although the ICP Global Office also developed a set of default values that were used when respondents could not or did not provide their own national resource mixes. The rationale behind the resource mix approach is that materials represent the final product of construction activity and are common – that is, comparable – across countries. The labour and equipment inputs, on the other hand, are only used during construction works and the proportions of these in resource mixes represent local practice, technology, productivity and other factors. The key is that volume measures of materials are directly comparable across countries whereas labour and equipment are country-specific and their volume or value depends on how construction is carried out in each country.

The research base for resource mixes is limited; an initial ICP note cited only around ten sources of data (Meikle 2011a). The initial estimates were, therefore, prepared on the basis of rather limited data although they were subsequently adjusted in the light of mixes received from countries during the ICP construction survey (Meikle 2013). Country responses, like the main survey price data, mostly came from single observers, although as noted previously, such observers bring broad experience to the exercise. Table 9.2 sets out both the initial and subsequently revised 'default' sets of resource mixes. The second set is based on responses from around 100 countries.

The major differences between the initial and the revised default mixes for residential and non-residential building work are a significant increase in equipment percentages of around 100%, a smaller (10%–20%) increase in labour percentages and a 10%–15% decrease in material percentages. Civil engineering

Table 9.2 Initial resource mixes and possible adjusted mixes (Meikle 2011a, 2013)

Groups of countries	Residential			Non-residential			Civil engineering		
	Mat.	Equip.	Lab.	Mat.	Equip.	Lab.	Mat.	Equip.	Lab.
Initial Global Office averages									
Low-income countries	72.50	7.50	20.00	72.50	10.00	17.50	50.00	35.00	15.00
Middle-income countries	72.50	5.00	22.50	70.00	7.50	22.50	50.00	28.75	21.25
High-income countries	70.00	5.00	25.00	66.67	7.50	25.83	50.00	25.00	25.00
Revised default mixes									
Low-income countries	62.50	15.00	22.50	62.50	17.50	20.00	No change		
Middle-income countries	60.00	12.50	27.50	60.00	15.00	25.00			
High-income countries	57.50	10.00	32.50	57.50	12.50	30.00			

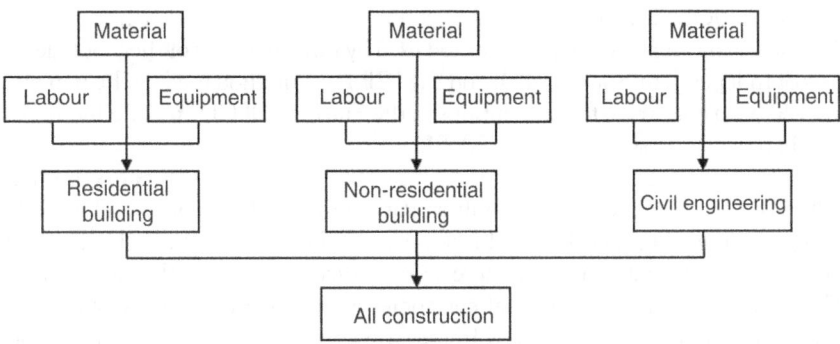

Figure 9.1 Construction PPPs for resources, basic headings and all construction

mixes are much more variable than building mixes, but the initial averages were broadly similar. Overall there was broad agreement among countries on resource mixes, but more work is needed on this aspect.

Other weights, from basic heading PPPs to 'all construction' PPPs, are more straightforward; simple expenditure value weights, provided by countries, were used. In summary, aggregate price relatives were used to obtain resource PPPs, resource mixes are used to obtain type of work (basic heading) PPPs and expenditure weights are used to obtain 'all construction' PPPs.

As indicated in Figure 9.1, the resource approach to the calculation of PPPs adopted in ICP 2011 generates 13 separate PPPs: labour, material, equipment and aggregate PPPs for residential, non-residential and civil engineering construction as well as the aggregate 'all construction' PPP.

Only one, the 'all construction' PPP, is published for each country but basic heading-level PPPs are available to researchers through the ICP 2011 data access policy (World Bank 2012). Detailed study of all 13 PPPs might allow researchers to identify the main drivers of 'all construction' price levels: is it labour or materials, is it residential or civil engineering construction, or is it some other combination? It might also be possible to comment on other issues including, for example, productivity levels.

The ICP construction results

In addition to a detailed description and discussion of the survey, the published ICP results (World Bank 2015a) provide a detailed set of tables that includes almost 200 countries. The main tables are:

- GDP PPPs and PPPs for selected economic categories;
- PLIs for GDP and selected categories; and
- GDP and basic heading expenditures in national currencies in USD using exchange rates and in USD based on PPPs.

The ICP values for construction PPPs are based on three calculation methods: the Eurostat/OECD method used by 47 countries across Europe, North and South America, Asia and Oceania; the method used by nine members of the Confederation of Independent States (CIS); and the ICP 2011 method used by the remaining 120 or so countries; for descriptions of the Eurostat/OECD and CIS methods, see chapter 5 in Best and Meikle (2015).

The main purpose of the ICP is, of course, to produce GDP and sub-GDP PPPs. The ICP also presents expenditure data in national currencies and in USD using exchange rates and PPPs. PLIs are derived directly from PPPs, by dividing PPPs by exchange rates, and are expressed as indices with the world equal to 100. The expenditure data in national currencies is provided by national statistical offices. The ICP results include 13 categories of consumption (from food and drink and clothing through housing and transport to health and education) and the two main categories of investment, machinery and equipment (M&E) and construction.

Where possible, this chapter uses normalised data and a sample of countries. PLIs are normalised (comparable) representations of PPPs; expenditure data is normalised by dividing by national populations and expressed as expenditures per capita or as percentages of GDP. The sample of countries is used to illustrate and comment on the results; the main rationales for their selection are that the countries cover the main ICP regions or groups, that they are significant in terms of population and economic size in their region, and that they exclude countries where PPPs have been calculated using the Eurostat/OECD or CIS calculation methods. The countries selected are listed in Table 9.3. The ICP regions and groups of countries are Africa (50 countries), Asia and the Pacific (23), Commonwealth of Independent States (9), Latin America and the Caribbean (39), OECD – Eurostat (47), Pacific Islands (23), Singleton Countries (2), Western Asia (12) and Dual Participation (countries included in more than one region) (6).

Purchasing power parities (PPPs)

The construction PPPs presented in the ICP results are annual and national average expenditure weighted aggregates of residential, non-residential and civil engineering PPPs calculated from the ICP construction price surveys. They are not linked to any particular type of construction project or any particular location. ICP PPPs are transitive; like exchange rates, they can be rebased to any country and the relative relationships will remain constant. Table 9.3 presents population and World Bank income group data and exchange rates, GDP PPPs and investment PPPs for a range of countries relative to one USD.

With the exception of Hungary and Saudi Arabia, the richer upper-income countries have GDP PPPs at the same or slightly higher levels than their exchange rates; Hungary's and Saudi Arabia's GDP PPPs are markedly lower, probably because of relatively low costs of labour. The upper-middle-income and lower-middle-income countries all have GDP PPPs lower than their exchange

Table 9.3 Selected countries, key indicators (World Bank 2012)

Country	Region⁺	Population (millions)	GDP per capita USD PPP	Income Group*	Exchange rate to USD	GDP PPP (base USD = 1.00)	Investment PPPs	
							Construction	Machinery and equipment
Brazil	LAC	192.38	14,639	UM	1.673	1.471	0.722	2.823
China	Asia	1,341.98	10,057	UM	6.461	3.506	2.184	7.771
Colombia	LAC	47.09	11,360	UM	1,848.139	1,161.910	883.72	2,528.15
Costa Rica	LAC	4.59	13,030	UM	505.664	346.738	233.246	798.305
Hungary	E/OECD	9.97	22,413	H	200.966	133.650	102.368	209.985
India	Asia	1,215.96	4,735	LM	46.670	15.109	9.598	48.134
Indonesia	Asia	241.04	8,539	LM	8,770.433	3,606.566	1,920.377	9,087.622
Netherlands	E/OECD	16.69	43,150	H	0.719	0.832	0.690	0.920
Saudi Arabia	WA	28.38	48,163	H	3.750	1.837	0.876	3.279
South Africa	Africa	50.46	12,111	UM	7.261	4.774	2.782	9.138
Tunisia	Africa	10.59	10,319	UM	1.408	0.592	0.253	1.913
UK	E/OECD	62.74	35,091	H	0.624	0.698	0.546	0.668
USA	E/OECD	312.04	49,782	H	1.000	1.000	1.000	1.000

* World Bank income groups, 2011: H = high income (>USD PPP 12,276); UM = upper-middle income (USD PPP 3,976–12,275); LM = Lower-middle income (USD PPP 1,006–3,975).

⁺ LAC = Latin America and the Caribbean; E/OECD = Eurostat/OECD; WA = Western Asia.

rates. The different investment PPPs indicate more clearly the influence of PPPs on different components of economies. Machinery and equipment PPPs, representing internationally traded items, are often (unsurprisingly) close to exchange rates in most countries; construction PPPs, on the other hand, are much lower than exchange rates or GDP PPPs in poorer countries, reflecting the essentially local nature of construction. This indicates that construction expenditure and construction volumes tend to be understated, sometimes significantly so, in poorer countries.

Price level indices (PLIs)

PLIs are calculated by dividing PPPs by exchange rates. GDP, all economy, PLIs are calculated by dividing GDP PPPs by exchange rates; PLIs for parts of the economy are calculated by dividing, for example, construction, and machinery and equipment PPPs by exchange rates. Although PPPs are calculated for each component of GDP, there is only one commercial exchange rate; in the same way, there are PLIs for each component of GDP.

It may seem odd to bring exchange rates back into the discussion. The purpose of PLIs, however, is to allow price levels for different parts of the economy and whole economies to be compared across a range of countries. Just as exchange rates relate to a particular point in time, so do PPPs, so calculating the relationship between exchange rates and PPPs at that point in time is both valid and useful as PLIs can be directly compared where PPPs cannot. PLIs normalise PPPs and make them comparable. The PLIs published in the ICP 2011 results are presented as factors with the world equivalent to 100, and like PPPs, PLIs are transitive (they can be rebased to any country and the relationships between countries stay the same).

Table 9.4 includes the same countries as Table 9.3 and presents PLIs for the whole economy, construction, and machinery and equipment.

Table 9.4 Selected countries, PPPs (USD = 1.00) and price level indices (world = 100)

Country	GDP PPPs	Exchange rates	GDP PLIs	Construction PLIs	M & E PLIs
Brazil	1.36	2.43	113.4	88.0	144.3
China	3.45	8.19	70.0	68.9	102.8
Colombia	1081.95	2320.75	81.1	97.5	117.0
Hungary	128.51	199.47	79.3	103.8	89.3
India	14.67	44.10	41.7	41.9	88.2
Indonesia	3934.26	9704.74	53.0	44.6	88.6
Netherlands	0.90	0.80	149.1	195.5	109.4
Saudi Arabia	2.41	3.75	63.2	47.6	74.8
South Africa	3.87	6.36	84.8	78.1	107.6
Tunisia	0.58	1.30	54.2	36.6	116.2
UK	0.65	0.55	144.2	178.2	91.5
USA	1.00	1.00	129.0	203.9	85.5

Source: ICP 2011 results (World Bank 2015a).

Table 9.4 illustrates the value of PLIs; they can be compared directly in the table, unlike exchange rates or PPPs. For example, Brazil's GDP PLI (its general price level) is higher than that of Colombia but lower than that of the United States; its construction PLI is lower than that of Colombia and much lower than that of the United States. The table clearly shows that the range of machinery and equipment PLIs (85.5–144.3, 1.7:1) is much narrower than either GDP (41.7–149.1, 3.6:1) or, particularly, construction PLIs (41.9–203.9, 4.9:1). This supports the idea that price levels across countries for internationally traded items will tend to be closer to each other than those for more local items.

Construction expenditure data

In the ICP, construction expenditure data is provided to the ICP Global Office in national currencies by national statistical offices. It is (or should be) the gross value of construction output in each country's national accounts, that is, it should include all construction activity in the economy. It should, therefore, include all capital construction work (new work and major renovations or extensions) by construction contractors, by households and by others where their activity is registered to construction. There is evidence, however, that this is not necessarily the case or, at least, that what is included or excluded is not consistent across countries. Possible exclusions are discussed below.

Other chapters in this volume discuss the problems of measuring construction output data in the UK (see Chapter 4) and informal or shadow construction activity in Australia and New Zealand (see Chapter 5). These or similar problems occur in all countries and create issues of both measurement and comparability. According to a report by AT Kearney, in five major European economies (Germany, Italy, Poland, Spain and Turkey) construction has the most prevalent shadow economy of any sector, making up at least 30% of all work in that sector (AT Kearney 2013). It should be noted that the Kearney figures are based on modelled data, not survey data.

It should also be noted that ICP construction PPPs include professional fees although these are not (or not all, or always) included as construction output in many countries (e.g. in the UK, construction professional services are excluded from construction output data and included in the UK national accounts as professional services).

A recent survey of national statistical offices in Africa illustrates the variability in what is included in, or excluded from, construction in the national accounts of a sample of countries (Meikle 2011b). Table 9.5 sets out the range of inclusions and exclusions and the data collection methods.

The table demonstrates that the comparability of country construction data is questionable in a number of cases. More detailed information from countries collected at the same time as the survey indicates that historic survey data or estimates are updated using population or household growth, rates of urbanisation or consumption of construction materials, particularly cement, or some combination of these. Regular dedicated construction activity surveys are rare in Africa.

Table 9.5 Construction in the national accounts of selected African countries

Countries	Construction activity		
	By registered contractors	By unregistered contractors	By households
Botswana	Based on survey; very small work excluded	Excluded	Excluded
Ethiopia	Based on survey and estimates; very small work excluded	Based on survey and estimates	Based on survey and estimates
Malawi	Based on survey and estimates	Excluded	Estimated
Mauritius	Based on survey	Estimated	Estimated
South Africa	Based on survey	Based on survey	Based on survey
Swaziland	Based on survey	Based on survey and estimates	Urban buildings using modern materials included; otherwise excluded
Uganda	Estimated; repair and maintenance and very small work excluded	Estimated	Estimated

Source: Meikle (2011b).

Table 9.6 sets out GDP per capita and construction expenditure per capita data from the ICP 2011 results for the same set of countries, all in USD, using exchange rates and PPPs. Two methods have been used to aggregate individual country expenditures: Geary-Khamis (GK) and Gini-Elteto-Koves-Szulc (GEKS).

Detailed descriptions of the methods can be found in the *Comprehensive Report* of the 2011 International Comparison Program (World Bank 2015b: 255–256). In brief, the GEKS method is considered by many statisticians as superior but, as a result of using it, the components of GDP are not additive; it was used for the 2011 published ICP data. The GK data is additive and was used for ICP results up until the 1980s but is now considered statistically inferior for producing values for both GDP and the components of GDP Additive means that the components of GDP, including construction expenditure, sum to the total of GDP. The fact that the GEKS-based data does not allow the components of GDP to be summed to GDP means that the relationships between the components and between the components and GDP are not necessarily reliable. Experimental data using the GK method was provided to the author by the World Bank Global Office.

The GEKS data generally looks to be in line with expectations, that is generally, values increase in poorer countries and reduce in richer countries and the increases and decreases in construction expenditure are generally greater than those in GDP. The GK data, on the other hand, is less predictable with only Chinese, Indian, Saudi Arabian and Tunisian real values greater than their nominal values and Dutch, Hungarian, UK and US real values significantly lower.

Table 9.6 Nominal and real GDP and construction expenditure data in USD

Country	Nominal expenditure in USD bn using exchange rates		Real expenditure in USD bn using PPPs and GEKS weights		Real expenditure in USD bn using PPPs and GK weights	
	GDP	Construction	GDP	Construction	GDP	Construction
		Amount %		Amount %		Amount %
Brazil	2,476.6	197.7 8.0	2,818.3	458.2 16.3	1,380.1	166.2 12.0
China	7,321.9	2,106.3 28.8	13,495.9	6,230.3 46.2	7,514.8	2,474.6 32.9
Colombia	336.3	48.9 14.5	535.0	102.2 19.1	266.0	35.2 13.2
Costa Rica	41.0	4.4 10.7	50.8	9.5 18.7	29.4	3.6 12.2
Hungary	137.5	12.8 9.3	233.5	25.2 10.8	122.9	10.2 8.3
India	1,864.0	334.7 18.0	5,757.5	1,627.2 28.3	3,293.9	665.4 20.2
Indonesia	846.3	219.3 25.9	2,058.1	1,001.7 48.7	1,234.2	393.7 31.9
Netherlands	832.8	84.0 10.1	720.3	87.6 12.2	426.2	35.2 8.3
Saudi Arabia	669.5	75.2 11.2	1,366.7	322.0 23.6	774.6	129.1 16.7
South Africa	401.8	36.5 9.1	611.1	95.3 15.6	253.7	29.4 11.6
Tunisia	46.0	6.3 13.7	109.3	35.2 32.2	62.6	13.8 22.0
UK	2,461.8	202.3 8.2	2,201.4	231.4 10.5	1,175.8	95.1 8.1
USA	15,533.8	1,295.0 8.3	15,533.8	1,295.0 8.3	8,215.4	529.2 6.4

Source: ICP 2011 results (World Bank 2015a) and World Bank experimental data (unpublished).

The expenditure data in the ICP results presents at least two problems for construction analysts. First, the basic data provided by national statistical offices may not represent the same concepts; and second, the non-additivity of the published data in PPPs does not, for example, allow credible figures for construction expenditure as a proportion of GDP (or GFCF) to be calculated.

Summary and conclusions

A first and important conclusion is that, although the ICP results may not be absolutely reliable, they are much better than anything else available. The more reliable figures in the ICP results are probably the GDP and GDP PPPs data and some of the less reliable are data on the components of GDP, including the construction expenditure data. Greater awareness about the ICP and greater involvement in its work by all, including the construction industry and construction researchers, will help encourage and direct that improvement. Recent initiatives by the ICP suggest that a number of recommendations made below are being addressed by the Global Office (World Bank 2018b).

There are shortcomings in the ICP 2011 documentation and approach and these will have influenced survey outcomes. There is uncertainty in item selection, price collection and weights, all of which can impact on the quality of the results and, in combination, may compound any individual inaccuracies. The approach, however, is not fundamentally flawed; international construction price comparisons are just very difficult, and the difficulties should not be underestimated.

The survey documentation needs improvement and more training of respondents and more checking and validation of survey data is required. Almost certainly, too much time was spent in the run-up to ICP 2011 on construction PPP theory and methodology and not enough on practical processes and data quality. The following aspects of the construction survey deserve attention:

- The list of items and item descriptions and supporting information should be reviewed and revised where necessary. Changes could usefully be made to the choice of items and supporting documentation; to the treatment of alternative materials and products; and to the identification and adjustment of item units. As a first step, initial reviews of the survey should be revisited and updated (see, for example, Meikle and Thomas 2013).
- The collection of data on and the treatment of mark-ups, profits and productivity needs to be re-examined and new approaches developed.
- More effort is needed on the selection of national construction experts and their familiarisation with the purpose and content of the survey and survey documentation. There is great reliance on expert pricing and prices, and enough time and effort needs to be put into informing the experts about the survey and how it should be completed.
- Rigorous procedures for checking and validation are needed, and these need to involve national construction experts. Checking of prices and adjusting for alternative materials and alternative units provide opportunities for error, and enough time needs to be allowed to ensure that adjustments are made correctly and confirmed with respondents. Construction prices are highly variable and this needs to be recognised.
- Benchmark prices, i.e. prices from non-survey sources, were introduced in the 2011 survey as checks on, not alternatives to, respondents' prices. More work could be done on this, for example, using official national average construction earnings data or commodity price data. A recent survey by Chinganye and others indicated that a significant proportion of countries regularly collect price data on construction materials (around 50%) and labour (around 25%) (AfDB 2015).
- Research is needed on resource mixes. The ICP 2011 data is almost certainly the largest international exercise in collection of construction resource data to date. More work is required to test the reliability of this data.

It is important that the data collected in the construction survey is the best possible within realistic time and cost constraints. Good quality data is essential, regardless of the PPP calculation method adopted.

The availability and reliability of input-output tables is increasing all the time. Weights for inputs to different types of construction work and all construction based on analysis of input-output tables should be collected and used to produce alternative PPPs as a check against PPPs produced using unweighted price relatives.

The PPPs and the PLIs in the ICP 2011 results illustrate the broad principles of PPP theory: that general price levels in poorer countries are higher than suggested by commercial exchange rates and that price levels for locally produced products, including construction, are also higher. The result is actual quantities or volumes of construction work in poorer countries tend to be understated using exchange rates. But PPP and PLI data for individual countries is indicative only and relative differences in price levels between countries should not be taken as precise. Aggregate PPPs for 'all construction' are weighted averages of the three basic headings, and these can be distorted by PPP values for individual basic headings and by the mix of basic headings – they are not necessarily comparable across countries.

PPPs are calculated for individual resources, basic headings and all construction (13 in all; see Figure 9.1). Analysis of these PPPs can help explain price differences, and the reasons for these differences, in a way that single construction PPPs cannot. Basic heading-level PPPs, including construction PPPs, are available to researchers through the ICP 2011 data access policy (World Bank 2012).

It has been noted by a number of observers that the data and approaches used to produce PPPs could also be used to produce temporal price indices. All countries have difficulty in producing reliable indices of construction price changes over time as well as construction output deflators, and it seems sensible to investigate linking work on both spatial and temporal indices for construction. Again, this is a task for the construction community.

There are reservations about the reliability and comparability of construction expenditure data, particularly in less developed countries. In Chapter 5 of this book Chancellor *et al.* discuss problems with the shadow economy in construction in developed countries. It may be that shortcomings in the quality of construction expenditure data are more significant than any problems with construction PPPs. More engagement and work is needed from the construction research community on methods for the collection or estimation and analysis of data on construction activity.

In addition, the method of aggregating expenditures in PPPs in the ICP leads to amounts (e.g. for construction) that do not sum to GDP. While this may be acceptable to and even preferred by statisticians, it produces confusing results for construction analysts that need explanation.

The ICP construction results present one of the most complete international datasets for construction research. The focus of the ICP, however, is the production of PPPs for whole economies and broad components of GDP. The calls here for more, and more detailed, information on construction PPPs cannot realistically be addressed by the ICP Global Office; there is a need for the construction sector, including industry and academe, to take a lead in analysing and presenting more complete and industry-relevant data.

References and further reading

ADB (2018) *Input-Output Tables (IOT) of Selected Economies in Asia and the Pacific.* https://sdbs.adb.org/sdbs/jsp/ICP/IOTDownload.jsp.

AfDB (2015) The reliability of economic statistics in Africa, focusing on GDP measurement. *The African Statistical Journal,* **17**, December, 17–49. oads/afdb/Documents/Publications/African_Statistical_Journal_Vol.17_-_01_2015.pdf www.afdb.org/fileadmin/uploads/afdb/Documents/Publications/African_Statistical_Journal_Vol.17_-_01_2015.pdf.

Kearney, A. T. (2013) *The Shadow Economy in Europe*. www.atkearney.com/financial-services/article?/a/the-shadow-economy-in-europe-2013.

Best, R. (2008) *Development and Testing of a Purchasing Power Parity Method for Comparing Construction Costs Internationally* (Sydney: Unpublished PhD thesis, University of Technology). www.dace.nl/download/?id=14800697.

Best, R. and Meikle, J. (2015) *Measuring Construction: Prices, Output and Productivity* (London: Routledge).

Davis Langdon (n.d.) *Variability in Eurostat Construction Pricing*, Eurostat Working Paper, unpublished.

ICP (2011) *A New Approach to International Construction Price Comparison*. International Comparison Program: http://siteresources.worldbank.org/ICPINT/Resources/270056-1255977254560/6483625-1273849421891/110622_ICP-OM_Construction.pdf

Meikle, J. (2011a) *Resource Mixes for Construction*. ICP Working Paper, unpublished.

Meikle, J. (2011b) *Measuring Construction Activity in the National Accounts of African Countries*. ICP Working Paper, unpublished.

Meikle, J. (2013) *A Second Update of Construction Resource Mixes for the ICP 2011 Round*. ICP Working Paper, unpublished.

Meikle, J. and Thomas, P. (2013) *An Initial Commentary on the ICP Survey and the Construction Survey Form*, unpublished.

OECD (2007) *Glossary of Statistical Terms*. https://stats.oecd.org/glossary/about.asp.

OECD (2018) *Input-Output Tables*. OECD.Stat. https://stats.oecd.org/Index.aspx?DataSet Code=IOTS.

Sinclair, N., Artin, P. and Mulford, S. (2002) Construction cost data workbook. *Conference on the International Comparison Program*, World Bank, Washington, DC, March. (DMS International). www.scribd.com/document/216887572/Construction-Cost-Data-Workbook.

Thomas, P. (2013) *Variability in ICP 2011 Construction Pricing*. ICP Working Paper, unpublished.

United Nations (2007) *Handbook of the International Comparison Programme: Annex II – Methods of Aggregation*. http://unstats.un.org/unsd/methods/icp/ipc7_htm.htm.

World Bank (2012) *2011 ICP Data Access and Archiving Policy Guiding Principles and Procedures for Data Access*. International Comparison Program. http://siteresources.world bank.org/ICPINT/Resources/270056-1255977254560/121120_ICPDataAccessPrinciples& Procedures.pdf.

World Bank (2015a) *Results of the 2011 International Comparison Program* (Washington, DC: World Bank). http://siteresources.worldbank.org/ICPEXT/Resources/ICP_2011.html.

World Bank (2015b) *Purchasing Power Parities and the Real Size of World Economies: A Comprehensive Report of the 2011 International Comparison Program* (Washington, DC: World Bank). http://pubdocs.worldbank.org/en/711001503680105564/ICP2011-Global-Report.pdf.

World Bank (2015c) *Measuring the Real Size of the World Economy: The Framework, Methodology and Results of the International Comparison Program* (Washington, DC: World Bank).

World Bank (2015d) *Operational Guidelines and Procedures for Measuring the Real Size of the World Economy* (Washington, DC: World Bank). www.worldbank.org/en/programs/icp/brief/2011-operational-guidelines.

World Bank (n.d.) *Construction Material Catalogue*. ICP Operational Material http://css.escwa.org.lb/ICP/1676/Construction-Doc-MaterialsCatalogue.pdf.

World Bank (2016) *International Comparison Program: Governance Framework* (Washington, DC: World Bank). http://pubdocs.worldbank.org/en/255521487200449880/ICP-GB01-Doc-Governance-Framework-Final.pdf.

World Bank (2018a) *International Comparison Program (ICP)*. www.worldbank.org/en/programs/icp.

World Bank (2018b) *Research Agenda: PPPs for Construction*. www.worldbank.org/en/programs/icp/brief/ra09.

Chapter 9 Appendix

ICP 2011 survey items: variability in prices and importance in basic headings

Item	Brief description	CoV	Default indicators of importance		
			Residential	*Non-residential*	*Civil engineering*
Materials and products		**1.07**			
Aggregate for concrete	Clean, hard, strong crushed stone or gravel free of impurities and fine materials in sizes ranging from 9.5mm to 37.5mm in diameter.	0.59	x	x	x
Sand for concrete and mortar	Fine aggregate washed sharp sand	0.56	x	x	x
Softwood for carpentry	Sawn softwood sections for structural use pretreated (to national standards) e.g. 50mm × 100mm	0.69	x	x	x
Softwood for joinery	Dressed softwood sections for finishing e.g. 18mm × 120mm	0.81	x	x	
Exterior plywood	Exterior quality plywood 15.5mm thick in standard sheets	1.32	x	x	x
Interior plywood	Interior quality plywood 12mm thick in standard sheets	1.31	x	x	
Chipboard sheet	Interior quality chipboard 15mm thick in standard sheets	1.19	x	x	
Petrol/gasoline	Standard grade for use in motor vehicles	0.78	x	x	x
Diesel fuel	Diesel fuel for use in construction equipment	0.89	x	x	x
Oil paint	Oil-based paint suitable for top coat finishes to timber surfaces	0.98	x	x	

Item	Brief description	CoV	Default indicators of importance		
			Residential	Non-residential	Civil engineering
Emulsion paint	Water-based paint suitable for internal plaster surfaces	0.95	x	x	
Ordinary Portland cement	Ordinary Portland cement in bags or bulk delivery	0.52	x	x	x
Ready mix concrete	Typical common mix 1:2:4 cement:sand: 20–40mm aggregate, 20 N/mm²	0.64	x	x	x
Precast concrete slabs	Precast concrete paving slabs 600mm × 600mm × 50mm thick	0.51	x	x	
Common bricks	Ordinary clay bricks (suitable for render or plaster finish) e.g. 215mm × 100mm × 65mm thick (715 bricks/m³)	1.03	x	x	x
Facing bricks	Medium quality self-finished clay bricks for walling, e.g. 215mm × 100mm × 65mm thick (715 bricks/m³)	0.78	x	x	
Hollow concrete blocks	Hollow dense aggregate concrete blocks, 7 N/mm², e.g. 440mm × 215mm × 140mm thick (76 bricks/m³)	0.79	x	x	x
Solid concrete blocks	Solid dense aggregate concrete blocks, 7 N/mm², e.g. 440mm × 215mm × 140mm thick (76 bricks/m³)	0.90	x	x	x
Clay roof tiles	Clay plain smooth red machine-made or similar tiles per square metre of roof surface area, e.g. 265mm × 125mm tiles	0.73	x		
Concrete roof tiles	Concrete interlocking tiles per square metre of roof surface area e.g. 420mm × 330mm tiles	0.74	x		
Float/sheet glass	Standard plain glass, clear float, 4mm thick	2.80	x	x	

Item	Brief description	CoV	Default indicators of importance		
			Residential	Non-residential	Civil engineering
Double glazing units	Factory made hermetically sealed, medium sized units 0.5 to 2.0m² with 4mm glass, 12mm seal	0.92	x	x	
Ceramic wall tiles	152mm × 152mm × 5.5mm thick white or light coloured for medium quality domestic use	0.83	x	x	
Plasterboard	12.5mm paper-faced taper-edged plasterboard in standard sheets	1.37	x	x	
White wash hand basin	Average quality white vitreous china domestic wash hand basin for domestic use, wall hung (excluding taps, trap and pipework)	1.88	x	x	
High yield steel reinforcement	Reinforcing bars up to 16 mm diameter (excluding cutting and bending)	0.54	x	x	x
Mild steel reinforcement	Reinforcing bars up to 16mm diameter (excluding cutting and bending)	0.52	x	x	x
Structural steel sections	Mild steel I-beams approximately 150mm deep and approximately 19 kg/m	0.63	x	x	x
Sheet metal roofing	Twin skin roofing panel comprising colour coated steel or aluminium profiled sheeting outer layer, 100 mm insulation, internal liner sheet,	0.72	x	x	
Metal storage tank	Metal storage tank capacity 15m³, thickness of steel, 5mm, typical size, 3.75m × 2m × 2m	0.96	x	x	x
Cast iron drain pipe	150 mm diameter with mechanical coupling joints	1.58	x	x	x

Item	Brief description	CoV	Default indicators of importance		
			Residential	Non-residential	Civil engineering
Copper pipe	15mm copper pipe suitable for mains pressure water.	0.99	x	x	
Electric pump	Electric pump for pumping water, temperature range, 5–80°C, flow rate 10 litres/second, head pressure, 150 Pa	1.41		x	x
Electric fan	Electric exhaust fan for interior installation, flow rate, 1,000 litres/second, head pressure, 250 Pa	1.60		x	
Air-conditioning equipment	Air cooled liquid chiller, refrigerant 407C; reciprocating compressors; twin circuit; integral controls cooling load 400 kW	1.49	x	x	
Stand-by generator	Diesel generating set for stand-by use, three phase 24V DC, 250 KVA output	0.79		x	
Solar collector	PV solar panels peak output 650 W, supply panels only, typically 4.5m² total area	0.77	x	x	x
Electricity	Typical average commercial tariff	2.21	x	x	x
Construction equipment		**0.28**			
Wheeled loader and excavator	1.0m³ loader capacity, 2.35m wide shovel, 6.0m max. dig depth	0.29			
Tracked tractor	Crawler dozer 159 kW with 'U' blade	0.35			
Skid steer loader	Tipping load, 2,000 kg, travel speed, 11.1 km/hr	0.31			
Tandem vibrating roller	Self-propelled 5-tonne double vibratory	0.26			
Compact track loader	Rated operating capacity, 864 kg, travel speed, 11.4 km/hr	0.34			

Item	Brief description	CoV	Default indicators of importance		
			Residential	Non-residential	Civil engineering
Construction labour		**0.19**			
General (unskilled) labourers	Workers that undertake simple and routine tasks in support of activities performed by more skilled workers. They have usually received little or no formal training.	0.24			
Bricklayer	These workers have	0.13			
Plumber	received training in	0.14			
Carpenter	their trades comprising	0.11			
Structural steel worker	one or more of apprenticeships,	0.13			
Electrician	on-the-job training or	0.19			
Machine (equipment) operator	training in a technical college or similar institution.	0.36			

Editorial comment

For some years a number of quantity surveying firms, particularly those that have offices in various countries such as Davis Langdon (now part of AECOM) and Rider Levett Bucknall, have collected and published international construction cost data. Often this data forms part of a larger report that includes general information about construction and property market conditions and trends as well as cost data of various types. Some publish these documents and make them freely available; others produce their reports for sale to stakeholders such as developers, other quantity surveying firms and government agencies. Generally they provide costs for a variety of typical building types, expressed as cost per square metre of floor area as well as unit rates for some typical building activities, indicative labour rates, possibly some plant/equipment hire costs and/or other related costs.

Turner & Townsend (T&T) began conducting a survey of building costs across many of their worldwide offices in 2009. They have continued to do this on an annual basis since then; at the time of writing, the 2018 version has just been released. Over the years the scope of the survey has increased and the number of offices providing information has increased. In this chapter the leader of the T&T program, Gary Emmett, describes the survey and its evolution from a small number of countries in 2009 to the considerably broader survey of 2018 as well as the evolution of the methodology and of the published data that is extracted from the survey.

Since 2013 Gary has been working on the method with the co-author, Craig Langston from Bond University. Their collaboration has led to the development of a more comprehensive approach to international cost comparison that includes the calculation of a form of purchasing power parity as an alternative to the more common presentation of costs in either local currencies or local currencies and USD equivalents based on current exchange rates.

The development and refinement of the methodology employed by T&T is still, to some extent, a work in progress and some aspects of it are still being debated and/or are under review. The T&T experience highlights a fundamental concern which has an impact on any attempt at international cost comparisons, regardless of the method employed and that is that there is no 'correct' answer against which results can be compared. Further complications include a generally low level of understanding of the theory and use of purchasing power parities

(PPP), even amongst many construction cost professionals, and the lack of agreed terminology for many of the inputs and outputs from cost comparison exercises. Terms such as 'price relative' are either not well understood or are replaced with other terms; even finding a universal generic descriptor for PPPs is problematic with different authors and agencies using a variety of descriptors including coefficient, index, multiplier, deflator, factor, conversion factor and real exchange rate, to name a few.

There are differing views on subjects such as the relationship between efficiency and cost; is it, for example, reasonable to suggest when comparing costs between locations that higher construction costs in a place indicate that the industry there is less productive? Or is it right to suggest that differences in construction costs indicate differences in cost of living? Answers to such questions are seldom clear-cut; more usually the answer lies somewhere between 'yes' and 'no' and is prefaced by words such as 'Well, it depends'.

As the authors describe in this chapter, the introduction of a PPP-style factor in T&T's annual report was not an immediate success, as many users did not fully understand what it was that they were using, nor did they know how to use it or when it was more appropriate to use PPPs rather than simple exchange rates. Best (2013) analyzed a report produced by the Business Council of Australia (BCA) that was widely reported in Australia at the time it appeared. It used data from T&T's publication to compare infrastructure costs in the United States and Australia with AUD costs converted to USD using annual average exchange rates. Based on their results the BCA suggested that Australia was a considerably more expensive place to build and that the Australian industry was, therefore, less productive than its US counterpart. Using PPPs to convert costs, Best demonstrated that the cost comparison results were reversed and he further argued that higher costs, in any case, can be the result of many other factors, such as differences in client expectations, differences in building regulations and so on.

The T&T experience illustrates some of the difficulties of collecting consistent and reliable cost data from many locations, regardless of the methodology used to develop that data into comparative costs and PPPs. No doubt their methods will continue to evolve, but the narrative in the following chapter provides some useful insights into the processes involved and may help others in the future to avoid some of the pitfalls.

10 Comparative construction cost data for industry

A case study of Turner & Townsend's experience

Gary Emmett and Craig Langston

Introduction

Successful project management has long been characterised in terms of delivering projects on time, within budget and to the required standard of quality (Ebbesen and Hope 2013). There are other performance indicators as well, including risk management, innovation, stakeholder satisfaction, value for money, environmental impact, defect minimisation, conflict avoidance, team development and continuous process improvement (Toor and Ogunlana 2010). Nevertheless, project cost is normally a key success factor for projects and therefore features prominently in benchmarking exercises aimed at identifying best practice (Bryde and Robinson 2005; Tabish and Jha 2012). Benchmarking concerns drawing comparisons between projects; in the case of cost, benchmarking is complicated by differences in scope, quality standard, timing and location (Atkinson 1999).

Investigations of comparative project cost performance may involve domestic or international benchmarking. The latter introduces the additional issue of different currencies. The routine approach is to first convert all costs into a common currency, usually taken as the US dollar (USD), so that a direct comparison can be made. Most practitioners appear to follow this approach. Yet currency rates can be quite volatile. For example, the currency exchange rate between Australia and the United States was about AUD 1 = USD 0.50 in 2001 and AUD 1 = USD 1.08 in 2012 (Best 2012), By late 2017 one Australian dollar (AUD) was buying around USD 0.75, and price levels in Australia and the United States still remained much the same as they had been.

Purchasing power parity (PPP) is an alternative to currency conversion. The concept has been around since the 16th century but was developed into its modern form by Cassel (1918) and has been used by economists ever since. It assumes that, in the absence of transaction costs and official trade barriers, identical goods will have the same price in different markets when the prices are expressed in a given currency (Krugman *et al.* 2010). Where this doesn't occur, the conclusion is that different parts of the economies in different countries have different domestic purchasing power.

A recent example highlights the cost conversion problem as it applies to construction. The Business Council of Australia compared the performance of large infrastructure projects in Australia and the United States and concluded that Australia had become uncompetitive (BCA 2012). Their press release was repeated in the national media:

> Australia has become such a high-cost and low-productivity nation that resources projects are now 40 per cent more expensive to deliver here than in the US, jeopardising an investment boom that is crucial to propping up the national economy. Landmark research to be released today finds that, compared with the US, airports are 90 per cent more expensive to deliver, hospitals 62 per cent, shopping centres 43 per cent and schools 26 per cent.
>
> (*The Australian* 2012)

Included in the BCA report were data on cost/m^2 for airports, schools, shopping malls and hospitals in both countries obtained from a cost guide published by Turner & Townsend (T&T). The BCA study was criticised by some analysts (e.g. Best 2012, 2013) for ignoring the impact of purchasing power. The original study benchmarked Australian projects against the US Gulf States where the cost of construction materials and the cost of living were lower than in many other parts of the country, resulting in different levels of construction prices. The BCA's observations enjoyed wide media coverage and were used politically to call for sweeping reform and productivity improvement, causing much angst and protest within the Australian construction industry and perhaps some misplaced pride in the US construction industry.

The aim in this chapter is to investigate the issues that surround the production and use of project cost data from the perspective of practitioners. A case study of T&T's *International Construction Market Survey* (ICMS) is used to explore comparison methodologies use of local currencies, USD and PPPs, and the role that each plays in different contexts. Appropriate cost comparison is essential to properly evaluate the success of projects as well as their initial feasibility. While the construction industry is used here as the context, the principles discussed apply to all types of projects, regardless of sector.

Case study

This case study is largely based on the personal experience of the lead author, Gary Emmett. He was, and is, responsible for the production of T&T cost data globally.

A basic question underpins the annual T&T exercise, a question that is of interest to many governments and their agencies, particularly in developed countries:

> Why has this country become one of the most expensive places in the world to build?

Alternatively, it could be:

> How expensive is it to build in this country compared to others?

This sort of question was posed, not for the first time, after publication of T&T's *International Construction Cost Survey* (ICCS)[1] in 2012, which presented comparative construction costs in terms of local currencies and USD (Turner & Townsend 2012).

Inevitably when international construction costs are published, the question of most interest to the reader is where their country sits in the league table of costs. The 2012 report, which was a development of earlier versions, mainly due to a broadening of the scope of the study (Turner & Townsend 2010), showed Australian construction costs close to the highest for most categories of construction. Soon after publication T&T received calls from journalists, CEOs, politicians and lobby groups wanting to know why costs were so high. Typical questions included "What has happened to construction productivity and efficiency?" and "How is the construction sector going to find ways of getting costs blowouts under control?"

At the time a higher than usual volume of mining and energy construction work was underway. Australia has had several instances of construction upswings across the various sectors, including mining, oil and gas, property and infrastructure in recent years. Whenever there is an upswing, the cost of construction resources such as skilled trades labour, engineering professionals, machinery operators and building materials increases. The perception that costs were spiralling out of control partially due to high wages paid to fly-in fly-out (FIFO)[2] workers during a boom period have dissipated now that demand has declined (Sydney Morning Herald 2016).

However, there is much more to the story than supply and demand for construction resources and high wages. Comparative cost studies often overlook the exchange rate implicit in the analysis and the different price levels in the countries being examined. In 2012 the Australian dollar increased above parity with the USD (i.e. AUD 1.00 was worth more than USD 1.00; the nominal exchange rate peaked at around AUD 1.00 = USD 1.09). As all the comparisons in the ICCS were done by converting local currencies to USD, the higher Australian dollar artificially inflated Australian construction costs and made labour and materials look very expensive. In the second quarter of 2018, with the Australian dollar around USD 0.76, Australia comes out somewhere in the middle of the pack of developed countries when the same calculations are used. Table 10.1 simply illustrates the effect of exchange rates.

This example shows that between 2012 and 2016 the cost to build a regional hospital in Australia increased by 15%. That translates to an average of slightly less than 4% increase per annum over four years, which most quantity surveyors and construction estimators would consider to be a reasonable rate of cost growth.

Table 10.1 Hospital construction costs 2012 and 2016 (Turner & Townsend 2016)

Regional hospital cost per m²	2012	2016	
Exchange rate USD/AUD	1.03	0.72	
			% change
AUD	3,303	3,800	+15%
USD	3,400	2,750	−19%

Yet in USD terms costs were 19% lower in 2016 than they were in 2012. At the 2016 exchange rate the 2012 Australian costs would be USD 2,390, or only 70% of what was published at the time and Australia would not have appeared to be one of the most expensive countries in the world.

This simple method of comparing costs by converting them all to a single base currency at current exchange rates is the approach used by most large quantity surveying firms that publish construction cost data. Typically the larger firms have global multinational clients who think in terms of how much a similar facility would cost to build in different countries, and doing the comparison in a single currency appears to work well for them.

However, this method is also inevitably used as a simplistic way of comparing construction performance. If it costs USD 800 million in Country X to build a hospital and USD 1.5 billion in Country Y, then does that really mean Country X is more efficient at construction? Government bodies might be very interested in the answer to that question. The answer, of course, is 'no'. It depends on the cost of wages, other resources and the exchange rate as well as the comparison currency used. Country X could be a developing country with very low wages but also low-skilled, less productive labour. They simply use more labour on the job because it is cheap and makes sense locally. Country Y may have high cost labour but use a lot of plant and equipment. In comparison with its labour cost it may be more productive.

In 2016 Switzerland stood out as the country with the highest costs in the T&T survey as shown in Figure 10.1, which takes the average cost in USD of six common building types in 38 locations (Turner & Townsend 2016). These are high-rise apartments, central business district (CBD) high-rise prestige offices, large warehouse distribution centres, general hospitals, primary and secondary school buildings and large shopping centres including malls.

Switzerland was not included in the ICCS back in 2012. However, Switzerland's high exchange rate, high wages and salaries, and also its high cost of living placed Zurich in the top spot ahead of New York and London in 2016. There is

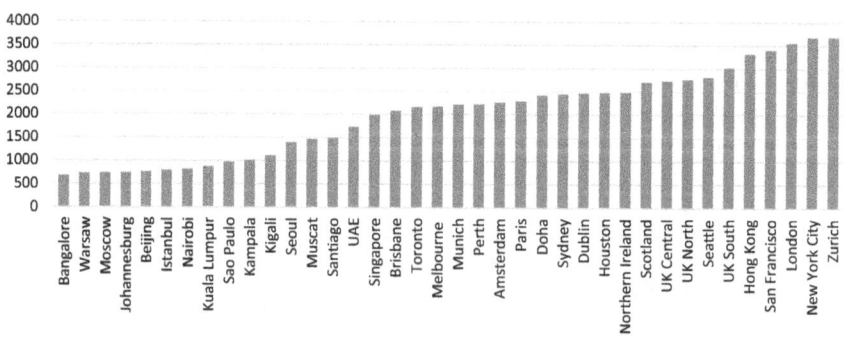

Figure 10.1 Average cost/m² for six building types in USD based on exchange rates (Turner & Townsend 2016)

another way of looking at this. Most of the world knows Switzerland as a high cost destination with high pay levels and a strong exchange rate. Swiss citizens are used to going overseas and buying cheaply because their exchange rate is so strong. The Swiss use a lot of technology and produce very high-quality buildings. To the Swiss, coming out as the most expensive place in the world to build is probably less of a concern than it is, for example, to Australians.

Following the 2012 survey report and numerous press articles T&T collaborated with Bond University's Centre for Comparative Construction Research (CCCR) to explore options for improving the presentation of comparative cost data in future survey reports.

There was a need for a way to demonstrate that comparing costs across national boundaries using money market exchange rates is less than ideal. Results can vary due to exchange rate fluctuations and conclusions that are based on such results about not only comparative costs but, by extension, industry productivity and efficiency, will often be questionable or completely false.

The CCCR suggested to T&T that a PPP method of comparison be used to help make up for the flaws inherent in the use of market exchange rates. Exchange rates often do not reflect the underlying cost of living as defined by the price levels in the local currency of various goods and price levels of particular activities in a country, and so may produce an over or under valuation of these activities. If the current exchange rate is USD 75 to AUD 100, it may well be the case that USD 75 would not purchase the equivalent in the United States that AUD 100 would purchase in Australia. Other factors such as the supply and demand for a currency are important. For example, a high demand for AUD by overseas companies wishing to purchase Australian iron ore was partly responsible for driving up the value of the Australian dollar in 2012. The underlying value of the currency was of less importance than the demand for currency at that time.

A PPP model may, for example, take into account the price levels in a country by relating costs to that of a basket of goods and services in local currency. What was needed from T&T's perspective was a simple way to introduce this into future publications to supplement the cost data already provided.

The CCCR supplied T&T with a new purchasing power parity model based on the price, in local currency, of a simple basket of construction commodities (resources) in a given location (i.e. city), which they called a *citiBLOC*. The concept of a citiBLOC involves selecting a representative basket of construction goods and services, not only in a given city but across the world, and pricing them in local currencies (Langston 2015a, 2015b). The relative costs of the basket of goods between regions becomes the PPP factor or multiplier that can be applied to each country's costs of construction in local terms. Rather than being measured in a particular currency, the cost of a building project in a city is now expressed as the number of citiBLOCs per square metre. Similarly, the cost of any construction goods or services can be expressed as the number of citiBLOCs needed to buy them. International comparisons therefore are made in terms of citiBLOCs (Langston 2015a, 2015b).

T&T prepares its reports by asking each region around the world in which it has offices to answer a survey that includes costs of approximately 50 common building materials plus construction labour costs, trade rates (the installed unit cost of materials) and total building costs for typical building types (school, high-rise apartment, etc.) expressed as cost per square metre. The survey, which commenced in 2009, is published regularly. The 2018 edition covers 46 locations, mainly cities, worldwide (Turner & Townsend 2018). Given the number of base prices in the survey, it was a simple process to price the citiBLOC basket for each city as all the required items were present.

The basket selected included the cost for ten items, commonly representative of construction in all regions, such as concrete, steel, glass, labour and plant. Quantities were set so that cost multiplied by quantity gave the cost of the basket components roughly in the ratio of 10% plant, 40% labour and 50% materials, which is no more than a rule of thumb.[3] Once the cost of the basket was determined, the basket costs in local currency were compared with a base city to obtain a PPP factor. The base in the most recent ICMS was set as London, although a better choice would probably have been for a city more in the middle of the international rankings, such as Sydney, which had been used previously (Turner & Townsend 2013). This change was made for political reasons given that London is the effective corporate headquarters for T&T, although arithmetically it makes no difference to the price relativities of location. The latest pricing is shown in Table 10.2.

The same quantities were then applied to the cost in each region's basket. So for example, the average cost of the Sydney basket was AUD 19,138 in 2016.

Table 10.2 citiBLOC basket pricing for London (Turner & Townsend 2016)

LONDON	citiBLOC calculation			
Basket item	Unit of measure	Cost/unit	Implied quantity	Total (GBP)
Labour:				
Electrician	hour	38	263	10,000
Carpenter	hour	35	285	10,000
Painter	hour	32	312	10,000
Unskilled labourer	hour	22	454	10,000
Material:				
32 MPa ready mixed concrete	m³	100	100	10,000
Reinforcement bar 16mm	tonne	860	11	10,000
10mm clear tempered glass	m²	100	100	10,000
100mm × 50mm sawn softwood stud	m	3	3333	10,000
13mm thick gypsum plasterboard	m²	5	2000	10,000
Plant:				
50-tonne mobile crane hire	day	1,200	8	10,000
			Average:	10,000

Comparing with a London PPP of 1.00 (GBP 10,000 divided by 10,000), the Sydney PPP in the 2016 ICMS is 1.91 (19,138/10,000). This exercise was repeated for each location in the survey to provide a PPP for each.

What was the benefit of this? We had an adjustment factor that we could apply to local costs that would show how much a specific type of building would cost in terms of the base location construction price level. To do this the cost/m² rate is divided by the PPP. An example is provided in Table 10.3.

Brazil experienced a dramatic currency devaluation following the slowdown in demand for Brazilian natural resources, largely as a result of slowing Chinese growth that significantly impacted its economy. Compared with a European economy, such as Ireland's, there was a dramatic difference in construction cost for the example selected (CBD offices – high-rise prestige). On a USD currency conversion basis, Dublin is 123% more expensive than Sao Paulo. It is easy to envisage businessmen and government officials being concerned at such a differential. Yet when we take into account the relative price levels in Brazil, by adjusting for purchasing power using the PPP factor, Dublin is only 3% more expensive. If we were making judgements about the relative construction price levels between Dublin and Sao Paulo, the PPP method of calculation would indicate that they are roughly similar – a better proposition than concluding that construction in Dublin was vastly more expensive, and therefore, by inference, rather less efficient, which is the conclusion we might reach if we used USD conversion at current exchange rates. Table 10.4 provides a further example.

In this example, it appears to be cheaper to build high-rise apartments in Singapore than in Sydney on a USD conversion basis. On a PPP basis Singapore actually comes out 29% more expensive. In other words, taking account of relative price levels, in Singapore it takes more construction resources (or citiBLOCs) to

Table 10.3 Example of PPP conversion for Sao Paulo and Dublin (Turner & Townsend 2016)

CBD offices prestige	Sao Paulo		Dublin		Difference
Local currency m² rate	6,000	BRL	3,050	EUR	
USD exchange rate conversion	1,490	USD	3,320	USD	123%
PPP factor	2.26		1.12		
PPP m² rate	2,655		2,723		3%

Table 10.4 Example of PPP conversion between Sydney and Singapore (Turner & Townsend 2016)

Apartments high-rise	Sydney		Singapore		Difference
Local currency m² rate	2,850	AUD	2,579	SGD	
USD exchange rate conversion	2,070	USD	1,860	USD	−10%
PPP factor	1.84		1.29		
PPP m² rate	1,549		1,999		29%

build high-rise apartments. This might be explained in several ways. Singapore is an extremely crowded city, and high-rise apartment construction might involve more expensive demolition, services relocations and traffic management than a similar project in Sydney. Other reasons are that labour is cheaper in Singapore relative to other local costs, and Sydney construction is perhaps more machinery intensive than Singapore.

There may be multiple reasons for variations in cost on a PPP basis. Variations in quality, specifications and inclusions in the standard product could also be an explanatory factor. At the least it goes a long way towards explaining why simple single currency conversion is not the best method for inter-city cost comparisons. Obviously, it would also be inappropriate to use one method in one context and another method elsewhere for a single study or publication.

So, armed with this new method of comparison, T&T started to introduce PPP and citiBLOCs into its annual cost surveys. The next section focuses on discussing our experience with this additional comparison method in the business world, as part of a quantity surveying practice.

Discussion

Key requirements of T&T's ICMS are that it needs to be:

- simple and easy to calculate; and
- easily understood by the intended audience.

T&T already had the cost data collected in the survey and, using the CCCR method of citiBLOCs, calculating the PPP factors[4] was quite straightforward. The second criterion, however, presented some challenges. Initially all of the PPP values were published alongside the local currency value and USD conversions at current exchange rate value (Turner & Townsend 2013, 2015, 2016, 2017, 2018). In the example presented in Table 10.3, three published rates for 'CBD offices – high-rise prestige' were given (BRL 6000, USD 1,490 and PPP 2,650). The Dublin equivalents were EUR 3,050, USD 3,320 and PPP 2,723.

This is where misunderstandings occurred. Many of the readers of the report use the data to make early stage assessments of where to build. If a CBD office building is being considered in Dublin or Sao Paolo, the USD values make it look very attractive to build in Sao Paulo. Using PPP narrows the gap considerably. Under PPP the difference is small enough (2,723 vs. 2,650) to make Dublin appear to be a reasonable option, especially if cost is not the only consideration. But if funds are being raised on the open market and repaid in the home currency, then the home currency value is what matters to the eventual owners; there is no currency called a PPP or citiBLOC. The PPP conversion is not what will actually be paid but is rather an index used for the purposes of comparison with other countries.

This is hard to explain to people in the field. Eyes glaze over and the concept is abandoned. Yet the practice of making judgements about whether the construction sector in a city delivers value for money based simply on cost/m² rates is a common practice, and PPP (or citiBLOC) is a valuable tool if such judgements are being made.

Another issue worth discussing here relates to the collection of cost data. Gathering, comparing and presenting construction cost data is a challenge. Having started out as a quick exercise in which 11 regional offices were canvassed for cost data, by 2016 it was based on a survey of staff in 38 of T&T's regional offices.

In each region, the 50 individual pieces of data included cost/m² rates for various common building types in residential, commercial, retail, industrial and civic construction. Common building materials costs, trade rates (i.e. labour and materials delivered and installed), and trades labour rates ('all-in' excluding overheads and margins) were also collected. Data on cost inflation, market outlook, typical margins and preliminaries costs are also collected. Because we are mostly interested in the building cost, local sales taxes, land costs, consultants' fees, local authority fees, loose furniture and fittings are excluded.

An issue arises because differences between building styles, methods of construction and inclusions affect the cost. When we first started producing the ICCS/ICMS we had to make a decision whether or not to have each region price single building designs with a bill of quantities. After all, this seems like an obvious way to get a true comparison of the differences in construction cost between countries. This idea was, however, quickly quashed: "But we don't build like that here" was a frequent response.

For example, in London, it is common to have a number of services in the floors of office buildings with both raised floors and suspended ceilings in typical use. In Australia, the services are usually in the suspended ceiling and raised floors are limited to IT server rooms. This affects the vertical distance between floors and hence the overall building geometry can be different for what might be common forms of building.

To get around this we ask the respondents to provide unit rates for the typical form of construction for that building type (school, multi-storey carpark, apartment tower, etc.) in their region according to local building standards, methods and typical inclusions. Our respondents are senior-level quantity surveyors in each region and most will have a good idea of what is the typical cost of the various building types in their region. We ask the respondents to supply the price of what is typical, so a certain amount of subjectivity is built into the data, and no doubt considerable debate takes place in the regional offices.

Where there are good recent examples, unit rates are relatively easy to supply. Generally there are plenty of good examples of materials, trades and labour costs from recent tenders. For our principal audiences this is valuable information. No doubt more work can be done in terms of understanding the cost differences of a higher quality build compared with a lower quality build. Another degree of subjectivity occurs because not every building type is constructed in each region or the sample size of observed buildings may only be small. Once again, we rely on the quantity surveyor to exercise a degree of professional judgement, and debate it with other experienced colleagues.

We have found it useful to compare all cost categories side by side and then to ask respondents to defend or amend anomalies. Using a coefficient of variation across the whole population helps with this process. Despite its complexity, our approach to data collection is practical and efficient, pushing against the limits of what we can practically ask staff already busily engaged on client projects to do.

Conclusion

After seven years of producing the international cost surveys, we have ended up with three ways of comparing information by location. No doubt further refinements are possible. Introducing ranges of costs to account for a spectrum of low to high quality may be useful. Experimenting with more items in the citiBLOC basket is another possibility. The citiBLOC items we have chosen may not reflect the higher and growing proportion of mechanical, electrical and services in newer buildings and may need to be reviewed in time. Other enhancements under consideration include an online app enabling easy comparison between regions at different points in time by applying escalation indices.

Importantly we now have more tools, and this proves especially useful when the debate turns to "Why has this country become the most expensive in the world to build?"

Notes

1 Turner & Townsend's annual publication was originally titled *International Construction Cost Survey*; since 2015 it has been titled *International Construction Market Survey*.
2 Fly-in, fly-out (FIFO) refers to a situation common in Australia where workers are transported to remote work sites where they work for a period of time (e.g. 20 days), usually every day, after which they are transported back to their homes for a week or so of downtime. During the recent mining boom in Australia which peaked around 2012–2014, FIFO workers were earning up to double the wages and salaries paid to people in similar jobs in the cities. These high pay rates combined with, among other things, the extra cost of transport of materials, provision of accommodation for workers, and the usual increase in tender prices that occurs when the construction industry is busy and work is plentiful, led to sharp increases in project costs.
3 Different agencies use different ratios, and ratios vary according to the type of construction. For example, civil engineering construction tends to utilises more plant and equipment than typical building work. Note also that the ratios are of *value*, not *quantity*.
4 Editors' note: T&T use the term *coefficient*, but strictly speaking PPPs are *factors* or *multipliers*. The World Bank calls them *indices*. The term *factors* has been used in this chapter for consistency.

References and further reading

Atkinson, R. (1999) Project management: Cost, time and quality, two best guesses and a phenomenon, it's time to accept other success criteria. *International Journal of Project Management*, **17** (6), 337–342.

The Australian (2012) Local project costs 40pc above the US, says Business Council of Australia. 7 June. www.theaustralian.com.au/national-affairs/local-project-costs-40pc-above-the-us/story-fn59niix-1226386836012.

BCA (2012) *Pipeline or Pipe Dream? Securing Australia's Investment Future*. Business Council of Australia. www.bca.com.au/Content/101987.aspx.

Best, R. (2012) International comparisons of cost and productivity in construction: A bad example. *Australasian Journal of Construction Economics and Building*, **12** (3), 82–88.

Best, R. (2013) Comparing project costs internationally: Methodology and data issues. In: *Proceedings of 27th IPMA World Congress*, Dubrovnik, September–October.

Bryde, D. J. and Robinson, L. (2005) Client versus contractor perspectives on project success criteria. *International Journal of Project Management*, **23** (8), 622–629.

Cassel, G. (1918) Abnormal deviations in international exchanges. *The Economic Journal*, **28** (112), 413–415.

Ebbesen, J. B. and Hope, A. J. (2013) Re-imagining the iron triangle: Embedding sustainability into project constraints. *PM World Journal*, **2** (3). http://pmworldjournal.net/article/re-imagining-the-iron-triangle-embedding-sustainability-into-project-constraints/.

Krugman, P., Obstfeld, M. and Meiltz, M. (2010) *International Economics*, 9th ed. (Boston: Pearson Higher Ed).

Langston, C. (2015a) Performance measures for construction. In: Best, R. and Meikle, J. (eds.) *Measuring Construction: Prices, Output and Productivity* (London: Routledge), 157–182.

Langston, C. (2015b) Refining the citiBLOC index. In: Best, R. and Meikle, J. (eds.) *Measuring Construction: Prices, Output and Productivity* (London: Routledge), 183–204.

Sydney Morning Herald (2016) Pampered FIFO workers come back to earth with a thump, 8 July, www.smh.com.au/business/companies/fifo-workers-face-leaner-times-as-down turn-eats-away-their-perks-20160523-gp164q.html.

Tabish, S. Z. S. and Jha, K. N. (2012) Success traits for a construction project. *Journal of Construction Engineering and Management*, **138** (10), 1131–1138.

Toor, S. and Ogunlana, S. O. (2010) Beyond the 'iron triangle': Stakeholder perception of key performance indicators (KPIs) for large-scale public sector development projects. *International Journal of Project Management*, **28** (3), 228–236.

Turner & Townsend (2010) *International Construction Cost Survey 2010*. Turner & Townsend. www.turnerandtownsend.com/media/1604/international-construction-cost-survey-2010.pdf/.

Turner & Townsend (2012) *International Construction Cost Survey 2012*. Turner & Townsend. www.turnerandtownsend.com/en/insights/international-construction-cost-survey-2012/.

Turner & Townsend (2013) *International Construction Cost Survey 2013*. Turner & Townsend. www.turnerandtownsend.com/en/insights/international-construction-cost-survey-2013/.

Turner & Townsend (2015) *International Construction Market Survey 2015*. Turner & Townsend. www.turnerandtownsend.com/media/1603/international-construction-market-survey-2015.pdf.

Turner & Townsend (2016) *International Construction Market Survey 2016*. Turner & Townsend. www.turnerandtownsend.com/media/1518/international-construction-market-survey-2016.pdf.

Turner & Townsend (2017) *International Construction Market Survey 2018*. Turner & Townsend. www.turnerandtownsend.com/media/2412/international-construction-market-survey-2017-final.pdf.

Turner & Townsend (2018) *International Construction Market Survey 2018*. Turner & Townsend. www.turnerandtownsend.com/media/3352/international-construction-market-survey-2018.pdf.

Editorial comment

In 1943 the then president of IBM, Thomas Watson, is said to have made a prediction which is now famous for being spectacularly wrong. He reputedly said: 'I think there is a world market for maybe five computers'. Much more recently, IBM reported that 2.5 exabytes (2.5 billion gigabytes) of data was generated every day in 2012 (Wall 2014). By 2017 the prediction was that total global data would reach a staggering 163 zettabytes (i.e. one trillion gigabytes) by 2025 (Cave 2017). Much of that data is now being collected automatically by devices such as smartphones, mobile and fixed sensors of many types, security cameras and many more. The datasets that are the result of this massive data collection activity are so large and complex that they cannot be analysed or managed by the hardware and software that has evolved and served us well since the time of Thomas Watson's supposed prediction. With the rapid exponential growth in global data a new term, big data (or should it be Big Data? or 'big data'?), has emerged to describe these very large datasets and new techniques have been developed, and continue to develop, that make analysis of such datasets possible.

How big data is used is still very much a work in progress, although evidence of how it is used in some areas is clearly visible in our daily lives through phenomena such as targeted online advertising – for example, when we search the internet for a hotel room, magically we start to receive information about other hotels, holiday destinations and deals, car hire and other related topics. On a broader scale, whole populations, or at least large samples of populations, can potentially be influenced by targeted messages that are triggered by information gathered about many individuals based on websites visited, online advertisements clicked on and so on. There is strong evidence to suggest that parliamentary and presidential elections have been won and lost on the basis of how well big data was utilised to assess voter preferences and then to create political messages that swayed voters based on their personal characteristics, beliefs and preferences.

In this chapter, the authors provide some background on the evolution of big data and an overview of some of the ways in which big data is being used and may be used in the future. More specifically they explore some of the potential for the use of big data analytics in the construction industry. One example is the possible expansion of conventional measures of construction performance (e.g. measures based only on cost and time to build) with more wide-ranging measures that include parameters such as design and operating costs. In the past, collecting and processing data such as detailed energy usage from buildings has been difficult and expensive, if not impossible. In an age of cheap sensors, smart metering, energy loggers and building management systems that record water and energy consumption (and production, in buildings that generate their own electricity through solar cells and the like), there is far greater scope for assessing and comparing building performance and thus for benchmarking and performance improvement.

Big data is a relatively new concept and there is clearly plenty of potential for its application across many industries, including construction as well as related fields such as real estate and facilities management.

11 Applications of big data to construction

D'Maris Coffman and John Kelsey

Introduction

One of the authors (a 65-year-old man) had a debit card stopped twice in 2016 when attempting to buy a suit at a very well-known UK store. Despite the fact that there had been no other recent big-ticket spending the bank stated each time that their system had detected a 'criminal pattern of spending'. On this basis there must be many more criminals in the UK than records show. It emerged later that the bank's detection program used machine learning tools employing big data and artificial intelligence (AI). This suggests that uncritical reliance on such tools may be questionable.

The promise of big data threatens to seduce us both by its abundance and its self-assured claims of what it can deliver. Big data threatens to present itself as a solution to every problem and may divert researchers from addressing more serious but less tractable problems that require more relevant data which is perhaps harder to come by (Borgman 2016). The construction industry does not make optimal use of the data it already has because of poor systems of, or cultural barriers to, knowledge management (Egbu and Robinson 2005; Wei and Miraglia 2017) and weak communications (Emmitt and Gorse 2003; Jaffar *et al.* 2011; Olanrewaju *et al.* 2017).

In order to set out the case for the judicious use of big data in construction measurement, we pose and answer the following questions, the first two of which are principally addressed by other chapters in this book and its companion (Best and Meikle 2015).

- What types of problem are posed by the need to measure the economic value of construction?
- Where is data most obviously deficient in terms of quantity and/or quality?
- What is 'big data' and what data is, or may become, available which can act as a useful surrogate for current data with questionable quality and varying availability?
- What legitimate inferences may be drawn from apparent correlations between 'big' data sets and construction variables?
- What additional future construction economic statistics may be required by society in a world of constrained resources and climate change?
- What other technologies are coming into use which, when combined with big data analytics, might assist in providing additional useful data?

- How might big data be used to forecast future construction activity?

Problems of economic measurement in construction

The chief challenges faced by those seeking to measure economic activity within the construction industry may be summarised as follows:

- capturing the whole economic value of construction within fragmented systems of industrial classification in official national accounting systems
- justifying, in both private and public-sector organisations, the choice of recording the value of constructed assets, with due regard to the complication that most constructed assets are attached to a parcel of land, in 'Statements of Financial Position' (formerly known as Balance Sheets) using one of the three accepted asset accounting measures:

  - Capital Value at Cost
  - Discounted Value in Use
  - Market Value in Exchange (Net Realisable Value)

- comparing the value of constructed assets in terms of other goods and services,
- making inter-temporal and international comparisons of economic activity in construction.
- making comparisons between assets involving products of different quality or specification which fulfil the same purposes – housing being an obvious example,
- making comparisons between processes involving production methods of differing factor inputs or process locations which result in the same or similar outputs (e.g. differences in production technology).

Problems of missing and/or reliable data in construction

The deficiencies of construction data in relation to these issues is well documented (see, for example, Best and Meikle 2015). These can be summarised as follows:

- The data does not exist
- The data exists in an extremely unstructured form
- The data is stored through non-digital media for a possibly very limited period
- The quality of the data is dubious
- Data relating to any one issue may be fragmented among different data holders
- The data is private (and/or commercially sensitive) and not accessible to outside researchers
- There are genuine conceptual problems in interpreting some of the existing data.

What is 'big data'?

If readers are looking for an agreed 'scientific definition' then they will be disappointed; the term has been so overused in recent years that it is in danger of being one of those terms that everyone uses but nobody agrees on its meaning.

In a broad sense, the phrase connotes enormous, complex data sets that are too large to be tractable using traditional statistical methods. Boton *et al.* (2015) cite the McKinsey Global Institute's definition that big data comprises 'data sets whose size is beyond the ability of typical database software tools to capture, store, manage, and analyse' (Manika and Chui 2011). They also cite Provost and Fawcett (2013), who define big data as comprising 'data sets that are too large for traditional data-processing systems and that therefore require new technologies'. This amounts in practice to the common-sense definition that big data is 'any data that cannot fit into an Excel spreadsheet' (Batty 2013), which, though amusing, is not sufficient for our purposes.

If big data is simply defined by the size of the data set or the difficulty in processing it using a given level of computing power, then what is classified as 'big data' will necessarily change over time. For instance, in the 1920s and 1930s, when financial econometrics and time-series analysis were first developed by the Working brothers at the Stanford Food Research Institute and later at the Coase Commission (which encouraged the application of these techniques to securities prices), researchers struggled to deal with daily price and volume data on about 12 agricultural commodities from the North American grain exchanges (Coffman 2015). Over the last 30 years, researchers have had access to real-time ticker tapes of trades made, which with the advent of high-frequency trading in the 21st century now occur every millisecond.

Many of the techniques developed by Holbrook Working (often credited with the 'random walk' hypothesis about securities prices) and his brother Elmer Joseph Working (who discovered the 'identification problem' in statistical inference) are still in use today in modern statistical science (Saleuddin and Coffman 2018). In particular, time-series analysis is still widely used today by those who wish to spot patterns or trends in large data sets. If big data has a claim to being categorically distinct, requiring a range of new computing tools, then that claim cannot rest on the volume of the data alone, nor can volume alone be used as a justification for abandoning traditional methods.

In business economics and management studies, 'big data' refers to the use of predictive analytics in the processing of transactional data sets with the aim of predicting behavioural outcomes or in identifying occult trends (Gandomi and Haider 2015: 137). In built environment disciplines, 'big data' further suggests an agenda of 'smart cities', with its attendant opportunities, challenges and pitfalls (Van De Wetering *et al.* 2016; Kitchin 2014a). The controversies around big data are numerous, but they can be collapsed around two key issues for the purpose of this discussion. First, there are both intellectual and practical objections raised that suggest that many of the tools used to analyse big data are deliberately agnostic about causal relationships. There is no attempt to ask 'why' a particular correlation exists, but rather to lever the fact of the current correlation to predict future correlations. To some, this is a welcome move away from formal statistical modelling, with its simplifying assumptions, towards a 'new empiricism', whereas others see it as an abandonment of the scientific method and even the 'end of theory' (Kitchin 2014b: 6). Second, analysis of big data often engenders privacy concerns, especially in the face of data integration, where different, ostensibly unrelated data sets are combined to yield often highly private insights about individuals (Al-Saggaf and Islam 2015).

At an extreme, there is a concern that the inferences drawn can be unjust and unwarranted (e.g. if Facebook predicts a couple will break up or if an individual is having an affair), or even self-fulfilling (e.g. if a credit rating agency anticipates a customer might have trouble making payments, and thus existing creditors raise their rates or slash credit lines). Existing legislation and regulation varies by jurisdiction, and there are as yet few global standards (Kshetri 2014). The rise of predictive analytics has provided grist for dystopian fantasies, such as that elaborated in the blockbuster film *Minority Report*.

However, there is a general consensus in the literature that big data is best assessed and characterised in three different dimensions, known as the 3Vs, which can be expanded to 4Vs or 5Vs depending on the level of nuance desired; these dimensions consist of volume (quantity), velocity (speed at which the data is generated) and variety (heterogeneity in type of data available), and can also include variability (consistency or inconsistency of the observation) and veracity (data quality) (Kitchin 2013, 2014a; Gandomi and Haider 2015). Each of these dimensions has different implications for data management and analysis and, as a consequence, are all well studied.

In this chapter, the term big data is taken to mean the employment of modern predictive and prescriptive analytics to harness the potential of 'high-volume, high-velocity, high-variety' data sets (Laney 2001; Gandomi and Haider 2015), which are generated as by-products of routine transactions in the construction industry.

Big data sources

When most people hear the term big data, they probably think of the data collected by leading internet firms. This includes e-commerce data (e.g. Netflix, Amazon, Google), human-generated mobile and social media communications and automated data communications, all supported by vast increases in storage capacity through new technology. In an effort to encourage innovation in big data analytics, a number of organisations such as Amazon and Google provide routine access to large public data sets.

Apart from traditional computing communications, the global increase in smartphone use generates a huge quantity of data every day. So, too, do an increasing number of sensors which are fitted to buildings, plant, vehicles and other assets, which can also, in certain cases, directly communicate with each other – a development referred to as the 'Internet of Things'.

At the other end of the data flow, new technology allows the splitting, distribution, storage and retrieval of very large data sets through cloud computing using multiple remote servers. This can include everything from data generated by building information modelling (BIM) systems to till receipts for leading wholesale and trade sale building supply firms. The data generated through these various means is not only of a large volume but also grows with incredible speed. Combined with that, the data is of huge variety both in nature and also in the degree of structure exhibited by the data sets so that the usefulness of data may require different analytical treatments – not least because the regularity or consistency of data capture may vary along with the accuracy of the data itself.

Big data analytics

Researchers can choose from a wide variety of tools and techniques available for the analysis and interpretation of big data sets, some of which were developed by academics working on their own or in partnership with firms. Many of these techniques are also used to analyse smaller, lower velocity, less variable data sets. Anyone with decent programming skills and access to the internet can find open source solutions freely available online. The challenge for state statistical agencies and firms alike is to possess the technical skills in-house to deploy them effectively.

In the last few years, given the commercial value of predictive and prescriptive analytics, firms have invested significant resources in developing in-house, bespoke techniques, including 'black box' solutions with proprietary code which they later commercialise. An early example of the wholly commercial product is the data analytics firm Palantir, founded in 2004 by Peter Thiel as a spin-off of the financial fraud analytics developed by PayPal and the trading platform used by his global macro hedge fund, Clarium Capital. Palantir offers two black box data mining solutions: Palantir Gotham for government use in fighting terrorism and Palantir Metropolis for financial services firms. Nearly 15 years later, hundreds of such products come to market each year, leading to trade sales of the most promising tools for data mining and data visualisation to the largest industry players.

Whether or not the researcher employs black box solutions or a range of well-established open source solutions, big data tools and analytics can be grouped into several approaches by the primary techniques used.

Data mining

Data mining uses a set of methods developed by statisticians and database managers to find patterns in large data sets. Examples include discovering and operationalising association rules (such as the existence of 'market baskets' where customers who buy sausages and bacon might commonly also buy eggs and orange juice, or those who buy bagels might also buy cream cheese and gravlax), supervised learning of classifications within a data set (segmentation of customer characteristics such as payment method, frequency, average spend, etc.) where the

categories are known *ex ante*, unsupervised cluster analysis (similar to classification analysis but without any *ex ante* assumptions about the categories to be
generated), and standard regression analysis.

A relevant example for the building trades might be the use of till receipts
from a firm, like the UK firm Wickes (see http://wickes.co.uk), that caters to do-
it-yourself (DIY) retail customers, tradesmen, and small and medium enterprises
(SMEs) working primarily on home renovation and repair projects. With some
such firms, those working in the building trades can be classified *ex ante* either by
purchasing on account or being enrolled in a program that offers a discount or
some kind of rebate for trade purchases. On other occasions, cluster analysis can
actually segment large data sets without any prior knowledge about the identity
of the customer (i.e. it can identify customer type based on the goods purchased).
Additional data mining can detect patterns in, for instance, the kinds of materials or tools purchased, the size of the transactions or additional payment methods
(if not on account) in order to predict business cycles.

Data integration

Data integration or data fusion involves splicing unrelated large data sets together
in order to learn how they interact. These can be ongoing or can take the form
of event studies, for instance, around the implementation of a new regulation.
One example might be combining data from building sensors with data from
weather forecast services to determine if end users are less likely to follow energy
conservation guidelines on extreme weather days. Another application could be
to combine building sensor data with incident reporting data to see if high-profile
health and safety campaigns, with attendant text messaging and social media
communications, are having the intended effect.

Another example might involve integrating firm data from Bureau van Dijk's
FAME (*Financial Analysis Made Easy*) database with transaction data from land
registries to predict business cycles. Data integration can promote optimisation
strategies, where the intention is less to measure the effect than to maximise or
minimise it in, for instance, health and safety campaigns. Data integration often
employs signal processing, which can help differentiate 'signals' (useful information) from background 'noise'.

Machine learning

Machine learning refers to applications of artificial intelligence research in
computer science which develop algorithms that enable computer programs to
evolve their behaviours in response to new data. Most of these strategies rely on
pattern recognition. Natural language processing is one of the most developed
areas within machine learning, and thus can be helpful for parsing free-form data
sets. Some of these use neural networks whereas others use tools from semantic
logic to conduct sentiment analysis. Being able to judge shifts in sentiment in
online communities, including social media or chatter amongst traders of a given

financial asset, can be very valuable commercially, in predicting, say, housing markets. These approaches have also been used, with varying degrees of success, to predict election results (Bermingham and Smeaton 2011).

Network analysis

Network analysis or social network analysis may evoke thoughts of analysing Facebook newsfeeds, but in technical terms applies to the characterisation of networked social structures in terms of nodes and ties. These can include online social networks but are also used to analyse real world supply chains, kinship relations, disease transmission and even scientific collaboration. Social network analysis does not depend on big data and, in fact, most implementations of this approach in social science research involve relatively small data sets. Yet there is considerable potential for development in this area; using data mining techniques, networks can be modelled without dependence on *ex ante* knowledge of the network, that is nodes and weak and strong links can be identified through recourse to the data.

Observation versus optimisation

Observation versus optimisation strategies necessarily entail some degree of trade-off. Each of the strategies detailed above can be used to describe data sets and generate forecasts based on predictive analytics. Recently, however, prescriptive analytics have also become the domain of big data through optimisation strategies that guide decision-making. Applications to construction at the firm level can include optimisation of construction schedules or reduction of construction waste, and many would argue that prescriptive analytics represent the most commercially valuable future for big data.

Just as cloud computing allows a distributed storage and retrieval of large data sets, so other software can be used to carry out distributed processing of that data so that large and complex analytical tasks can be performed without reference to individual computer capacity.

Although the computing power required may be large and results may appear complex, the principles of big data analytics are relatively straightforward and are enumerated below.

Bloom filters

A Bloom filter is a technique which uses probability theory to look for a needle in a data haystack (Bloom 1970). Essentially the question is asked: 'is an x-type object a member of the dataset T?' The building block of this technique is a 'hash function', which is rather more complicated than the example given below but the example illustrates the principle.

Suppose there is a data set of words and we want to know if the word *plumber* is present in the data set. Instead of searching for the string of letters *plumber*, we simply input *plu*. If the string *plu* is not in the data set then we also know that *plumber* is not in the data set. However, the query may produce entries which are not *plumber*, such as *plug* or *plumbline*. So the technique can definitely tell us if the search item is NOT in the data set but there is a small probability that it will return additional items as present when they are not the same as the full-length search item.

The combinatorial probabilities afforded by using a number of such reduced-length searches mean that a search for the full-length item can be conducted in a manner which is much more efficient in terms of computing power than if we simply put in the full-length term, although at the risk of a small number of 'false positives'.

Such techniques permit the efficient querying of very large data sets.

Hierarchical clustering

If we have a type *x* renter of commercial property with a specific set of characteristics and we want to know something about other behaviours of such renters, we can look at a data set of those *n* renters with sets of characteristics which are nearest to our subject. We can then use the known behaviours of those 'nearest neighbours' to predict the behaviour of the subject. (That, at least, is the theory.)

Clustering looks at whole data sets and attempts to segregate them into clusters based on relevant sets of characteristics. It is thus very good for marketing analysis not only to select groups for determining what are the goods or service characteristics that should determine future mass production, but also to isolate specific 'out' groups, which may benefit from niche products or services.

Structuring goes a step further by, in effect, creating a hierarchy of characteristics. For marketing purposes, if the most important characteristic is the age of a potential buyer, then consumer age will go to the top of the tree and data will first be sorted by age, then another characteristic is used to divide the data further and so on.

To illustrate how such a hierarchy be constructed, suppose we have 200 male students:

150 of them wear blue socks and 50 wear red socks.
120 of them prefer coffee to tea and the other 80 prefer tea to coffee.
Of those who wear blue socks, 90 prefer coffee to tea.

If we split them first by sock colour we get:

Group A   90 blue/coffee (75% of coffee drinkers)

    60 blue/tea (75%)

Group B   30 red/coffee (25%)

    20 red/tea (25%)

If we split them first by drink preference we get:

Group C   90 coffee/blue (60% of blue sock wearers)

  30 coffee/red (60%)

Group D   60 tea/blue (40%)

  20 tea/red (40%)

Therefore, splitting first by sock colour is a better predictor of drink preference than is splitting by drink preference a predictor of sock colour.

In this case we can say that the A/B grouping produces more structured data than the C/D grouping. This then decides the order in which data is split. Obviously with more than two characteristics it becomes much more complex. Essentially the hierarchy decision at each level becomes a predictive accuracy competition. At any branch point in the tree the characteristic which best predicts all the remaining characteristics is the winner and is then used to further sub-divide the data.

A sophisticated version of this process, known as a Classification and Regression Tree (CART), was developed in the 1980s (Breiman et al. 1984). From a construction point of view the technique is useful in that it can bypass missing data by omitting predictors with missing values. However, if there is too much missing data, there is always the danger that the technique may be building the wrong tree.

With more complex data sets much more sophisticated methods are required to build the hierarchy, but the general principle is clear enough.

Artificial neural networks (ANN)

The previous techniques are potentially directly comprehensible even if the mathematics is somewhat complex. With artificial neural networks (ANN) we move into a technique which has proved itself to be powerful but where ordinary humans cannot directly observe what happens inside the black box. There is a huge literature related to this area, but the underlying principle is as follows.

The technique evolved from a debate about how the physical neural system works to produce the perceptions, decisions and actions which characterise human behaviour. The brain receives a large number of stimuli every second; there is evidence that the way the body processes those stimuli has a very different structure and process from our conscious understanding of perception, thinking, reasoning and deciding. This research is set out in McClelland and Rumelhart (1986) and Rumelhart and McClelland (1986).

The system essentially has a set of inputs (which could be big data), which are processed and mediated by a hidden layer of nodes to produced output data. The output data is determined by activation values (between 0 and 1) attached to the inputs and weights attached to the network connections which either increase or inhibit the communication of data. Initial outputs may well be inaccurate or unusable. Therefore, the weights are adjusted until they produce useful information. The method of changing the weights is determined by a training algorithm which reacts to output deviations, either from observed reality or approach to an optimum point, depending upon the task in hand.

One of the areas where ANN has proved its worth is in pattern recognition (Russell and Norvig 2002). Clearly this is very important for forecasting purposes; however, there are some fundamental problems which need to be addressed.

Assumptions in econometrics and big data

The main quantitative analytic techniques used by economists come at the moment from econometrics.

There are already enough reservations about simply estimating an equation in the form:

$$y_t = \alpha + \beta x_t + \varepsilon_t$$

and the legitimate inferences that may be drawn from the results.

Even before performing analysis, the sources, collection methods, consistency and accuracy of the data would be questioned and evaluated. Even then, evidence would be sought that the data is a representative sample of the population of like data. A decision would be made as to whether the data required some form of pre-processing. The specification of the model would be evaluated based on the behaviour of the error term. Questions would be asked as to the extent to which the error term arose from genuine disturbance influences or merely from measurement error. Hypothesis and appropriate interval tests would be constructed. R^2 values would be appraised and tabulated. In more complex models, problems including autocorrelation, heteroskedasticity and multicollinearity would also have to be addressed (Kennedy 1998).

If similar rigour is to be applied to big data, then the analyst must have much more than the data itself. They must have access to understanding the sources, reliability, consistency and precise nature of the data. Without that, any patterns of correlation will have limited validity and usefulness for any form of forecasting. The question arises then whether those who collect and/or produce big data analytics will offer similar levels of rigour as those required by the econometrician.

Granger causality

Granger (1969) proposed a test to determine whether one particular time series is or is not useful in predicting another. It has long been understood that correlation does not equal causality, but Granger argued that it should be possible to test for 'predictive causality'. The problem is that, if the following two relationships are true,

$$y_t = \alpha + \beta x_t + \varepsilon_t$$
$$w_{t+1} = \alpha + \beta x_t + \varepsilon_t$$

then this may appear as a Granger causal relationship between y_t and w_{t+1}. If, at some point, the relationship between x_t and either y_t or w_{t+1} breaks down, then any forecast based on what appears to be a relationship between y_t and w_{t+1} will also break down. Granger was, himself, well aware of the possibilities of the existence of such problems (Granger and Newbold 1974).

Omitted causal variables

Where there are omitted variables, it may also be that some pattern of economic behaviour is path dependent where the paths are mapped by the omitted variable. This is understood, for instance, in psychology where behaviour may be governed in some complex way by several different and fluctuating conditions.

Figure 11.1 represents a three-dimensional graph with a possible origin at either X point. Suppose a forecast model is mis-specified to show a relationship between variable x and y which is described by the plane XXYY. In fact, at certain points in the X range another variable comes into play, described by the plane XXWW, and the most extreme possible path is described by the plane WWZZ. Under certain conditions, instead of moving along a smooth straight line the Y variable may experience a discontinuous rise or fall represented by the thick black arrows. This can be described as a 'cusp catastrophe' (one of a number of possible forms). Zeeman's work follows that of Thom (1972), although it should be understood that Thom's models were not designed as forecasting models but rather as ways of topologically modelling possible forms of complex reality (including social models). Indeed, he was generally opposed to merely discovering patterns on the basis

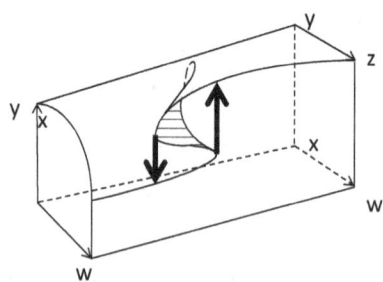

Figure 11.1 3D behaviour space with two independent control variables incorporating a cusp catastrophe

Source: Adapted from Zeeman (1976).

of predictive efficiency (Thom 1991). The key here is that the behaviour is path dependent according to whether movement starts at the left-hand or right-hand X point on the graph.

Hysteresis and real estate cycles

Another form of path dependency is where past behaviour governs current behaviour. Grenadier (1995) argues that real estate cycles persist because of demand uncertainty, adjustment costs and construction lags. Sellers of a number of goods and services will reduce their prices when there is reduced demand. However, this only happens very slowly in commercial property markets. Grenadier notes very considerable lags such that commercial landlords prefer to run with significant vacancies rather than offer units at more competitive rents. Similarly, landlords are observed to make significant concessions in terms of rent holidays or a lengthened period until the next rent review in order to keep existing tenants. They will, however, try to avoid lowering the stated rent on the lease.

The answer to this can be found partly in the costs of managing the transfer of a unit to vacant status as well as the cost of re-letting. Additionally, the exit of a tenant might send the wrong signal to the local market and/or other tenants in the same development. In retail developments the exit of a key tenant could have worse consequences. In terms of real options theory, a key piece of signal information is the headline rent per square metre. In the hope that the downturn is short and that rental values will recover, it is perfectly rational for landlords to choose to wait and incur losses rather than lower rents which they will be stuck with until the next series of rent reviews.

The other phenomenon Grenadier observed was that developers continued to overbuild well into the downturn even when this had become obvious. There is more than one lag involved here. First, there is the land purchase to development start lag; developers try to buy land when it is at its cheapest and develop when the market is more favourable. There is, however, a second lag at work in the form of the development start to completion lag. Additionally, there is a very considerable cost of abandoning construction once started. In parallel, there is a cyclical lag in demand such that the market may be in a different state when the development is complete. This makes for decisions which appear totally rational at the time but look foolish in hindsight. The combination of such individual decisions may then result in aggregate cycles with very pronounced fluctuations.

If one is using a real estate and construction demand model based on extrapolating past trends, this is clearly not going to work. However, no two cycles are the same and one cannot build a 'standard cycle' into the model either. While there is some evidence that adaptive forecasting models (employing machine learning tools) are better at, say, load forecasting over electricity grids, or predicting short-term box office receipts, standard econometric models still outperform machine learning tools at identifying long-run term trends (Coelho *et al.* 2016; Liu and Xie 2018).

The potential use of big data in different project phases

Project phases can be characterised in terms of economic value. The standard net present value (NPV) equation can be used to measure the underlying economic value of a building:

$$\sum_{t=0}^{n} \frac{(v_t - c_t)}{(1+r)^t} - K_0$$

and the profitability index can be used to measure the output/capital ratio or total productivity.

$$\sum_{t=0}^{n} \frac{(v_t - c_t)}{(1+r)^t} \times \frac{1}{K_0}$$

where n is the expected life of the building, v_t is the annual benefits or value provided by the building, c_t is the annual operation/maintenance cost of the building (including energy costs), r is an appropriate discount rate and K_0 is the initial capital cost. These expressions could also be used in the evaluation of a mid-life major refurbishment or indeed a mid-life evaluation of the building as it stands under 'technical due diligence' (Jensen and Varano 2011).

The initial capital cost can be decomposed into $K_0 = D_0 + C_0 + S_0$, where D_0, C_0 and S_0 are design, construction and sundry costs, respectively. K_0 can be considered as covering the design and construction phases while the remaining sum of discounted values represents the operating phase of the building. What, of course, the above omits is the actual market value of the property (including the building), which necessarily includes additional land-based, location-based, speculative and other elements.

One of the problems of construction performance measures is that they have tended to use only the initial construction cost (e.g. Langston 2015) ignoring both design and operating costs. Now this is entirely understandable in the context of (1) Standard Industrial Classification systems separating out design and construction and (2) lack of data relating building performance to initial capital (and more specifically construction) cost. However, this is clearly problematic. It is as if one attempted to measure the performance of the automotive industry without reference to the salaries of automotive design engineers or car running costs.

Yu and Ive (2008) suggest a hedonic model based on client willingness to pay for the construction of a building with some particular set of attributes. This certainly has potential for a more accurate means of economic comparison between different buildings, however, there are problems with even this approach as a longer-term solution.

The contribution of design labour to the value of a constructed product is not only obvious, it is enshrined in UK law in the form of the *Construction Design and Management Regulations 2015*, which provide that designers have to consider the long-term safety of their product in the operational phase and also the safety of the design in requiring certain processes to construct the building as designed. In other words, there is a design contribution (for better or worse) to construction and operational phase productivity.

So, there is a potential challenge in the extraction and analysis of data (hidden in many cases) in different parts of the wider architectural, engineering and construction (AEC) industries to establish the design contribution to construction output. Since the use of integrated forms of procurement (e.g. Design and Build, Engineer-Procure-Construct, Build-Own-Operate-Transfer) have greatly increased, an understanding of the relative contribution of both design and construction is clearly essential if those attempting to compare different constructed outputs are not to be reduced to studying a database of only those projects where the construction phase costs are easily identifiable (which may itself then constitute a biased sample). In terms of the operational phase, there is necessarily a degree of information asymmetry between the supply chain and the consumer, whether domestic or commercial. A much bigger problem, however, is the one of long-term risk and uncertainty regarding the performance and related operating costs of a constructed product (Capelhorn 2012).

Big data and emergent technologies

Among emergent technologies the following stand out as offering possibilities of being combined with big data to generate potentially useful construction measurement data:

- building information modelling (BIM);
- remote sensors, radio frequency identification (RFID), 3D printing and robotics;
- blockchain.

BIM has long been touted as a silver bullet that will remedy some of construction's problems. To some extent this is because different BIM software products are being, and have been, heavily promoted by software houses. There is nothing intrinsically wrong with this, and clearly the software producers have to be part of the solution, however, it is increasingly clear that a number of construction industry problems have to be solved by other means in order to make BIM work (Eastman *et al.* 2011; Alreshidi *et al.* 2017).

One of the authors, having practised as a quantity surveyor, knows only too well that collaboration combined with honest and regular data exchange is about the last thing to be expected in real construction contracts. In order for BIM to deliver its promise this is clearly necessary, and indeed the more progressive contractors do realise this and have shown some ability to change their ways. Additionally, there are as yet unresolved security, contractual and intellectual property rights issues about the ownership of specific types of data generated by construction activities.

Properly functioning BIM could deliver more tractable and structured data to enable better understanding of final account/tender price relationships, material wastage rates, labour/plant use analytics, energy consumption, cost/schedule overrun analytics and other specific forms of cost and risk analysis. There is, however, some way to go before this can be achieved.

The various new production technologies (e.g. sensors, RFID-enabled devices) already contain within them the means to transmit real-time data to other data capture devices or could be programmed to do so (e.g. 3D printing, robots). This allows construction-related activity data to be plugged into the Internet of Things for wider big data exploitation or to provide 'bigger data' for more specifically construction-oriented research. However, as with BIM, those holding the data must be prepared to share it in some form.

Blockchain is the technology behind the rise of e-currencies or crypto-currencies such as Bitcoin and it has been around somewhat longer than readers might think. While the actual technology is of interest, the real understanding needs to be conceptual. Trust has historically been in noticeably short supply in the construction industry, yet trust is at the heart of blockchain. The property rights that are taken for granted (whether owning homes, having security and privacy in bank accounts or enforcing construction contracts) only work because there is a degree of trust in some form of government to enforce those rights.

Those who did early development work on the internet realised that privacy and security were going to emerge as serious issues once the internet became a principal means of communication and business transactions (Tapscott and Tapscott 2016). One of the main reasons for international financial institutions such as the International Monetary Fund (IMF) and World Bank to have replaced gross national product (GNP) with gross national income (GNI) as the main measure of national income is the rapid rise in significance of internationally remitted earnings by migrant workers which forms one of the largest sources of international cash flows. However, they were often remitting to countries where banking and security were not as well developed as they might be. Therefore, they, along with other regular movers of cash, need a trustworthy system of international exchange and record. The 2008 global financial crisis dealt a blow to trust in financial institutions just as the 2001 collapse of Enron and WorldCom led to a crisis of confidence in corporate financial auditing and indeed led the largest auditing company in the world to cease to function.

The idea of blockchain is simple enough. The system is distributed among many servers throughout the world. It is, therefore, not under the control of any single government or institution. The ledger is public, in that anyone can access whatever they are authorised to access on the system as it resides on the internet. The system is heavily encrypted, so that privacy is at least as secure as bank data – and probably more so in practice. Data is assembled from transaction inputs into permanent records of time-stamped blocks which are then attached to preceding blocks which form a chained set of transaction histories (hence 'blockchain') (Tapscott and Tapscott 2016). Therefore, one cannot change any transaction without accessing and changing its entire history.

While this has been principally used for financial transaction data, there is no reason why it cannot be used for other data histories. This could entirely revolutionise construction procurement generally and public-sector procurement in particular. Governments could insist that all bids be delivered in BIM-structured and blockchain-enabled form so that there is an auditable record, not only of the bid sum but of all its component parts of design, quantities, warranties, costs and schedules including sub-contractor/supplier data and post-contract data of variations and agreed final account. There would exist, then, for researchers, the Office for National Statistics (ONS) and the National Audit Office (NAO), a set of records with which to measure and audit public-sector construction.

The tendency of government is to require more data to be supplied by individuals and companies. Given the impact of construction on the environment and in the production of carbon emissions, it would seem likely that government will require a greater quantity of better structured data relating to environmental sustainability in the future, which itself will feature more prominently in national accounts statistics.

Conclusion

While a comprehensive review of the uses of big data analytics in built environment studies is undoubtedly a larger topic than can be accommodated in a single chapter, the aim here has been to set out the terminology, correct a range of misconceptions about what big data can accomplish, and evaluate the relevance and performance of big data analytics for a range of facets of the measurement of construction.

Many sources of big data promise to yield significant new insights when properly analysed. The question remains how far machine learning tools, most often mobilised for the analysis of these data sets, represent an improvement over standard econometric models which often use smaller data sets to address some of the same questions. Although there is some evidence that they do so with respect to short-run forecasts, the main weakness of these methods is that they do not have a reasoned causal structure to support them as they are the result of model discovery. Thus, when they stop producing reliable forecasts, they must be discarded in favour of new models. There is no way to diagnose where they have gone astray; this is the basis upon which some authors have deemed them unscientific. Without wading into that debate, we would simply caution that such solutions may prove unsuitable for government statistical offices, insofar as they introduce real challenges of comparability over time and across space.

In terms of the construction industry, an especially exciting area for the application of big data analytics is in improving coordination between the design/briefing phase and the operational phase of a building's life cycle, and in providing evidence for estimation of whole life costs and benchmarking building performance. This is particularly the case because facilities management (FM), as part of the asset cycle, is seldom subject to the same optimisation efforts as the building phase of the project cycle. As more and more data is recorded in

BIM-structured databases, there is an opportunity to bridge that gap; attempts to do so, however, remain bedevilled by problems of interoperability and data integration (Pärn *et al.* 2017). Going forward, the construction industry will need to agree (ideally global) standards for data handling.

Another exciting frontier, beyond the scope of this chapter, which deals with present uses of big data in the measure of construction activity, is the potential for amalgamating BIM models along the supply chain in order to produce a financial and physical input-output model for each project. This should enable calculation of material and plant use for the whole building from which one could also calculate the embodied energy in the materials used. Combined with figures for final building occupant energy use, this would give us total consumption of scarce materials and the carbon output of buildings through material requirements, construction processes and occupant behaviour in use. While few would currently voluntarily provide such figures, it may in future be a regulatory requirement to do so as material shortages and energy costs both increase.

In closing, big data does offer new possibilities for the construction industry, but it also highlights the challenges facing the industry in the efficient use of data at any scale. This represents a greater challenge to all interested parties than the question of whether or not big data analytics, cautiously used, can yield new insights. That much, at least, seems clear.

References and further reading

Alreshidi, E., Mourshed, M. and Rezgui, Y. (2017) Factors for effective BIM governance. *Journal of Building Engineering*, **10**, 89–101.

Al-Saggaf, Y. and Islam, M. Z. (2015) Data mining and privacy of social network sites' users: Implications of the data mining problem *Science and Engineering Ethics*, **21** (4), 941–966.

Batty, M. (2013) Big data, smart cities and city planning. *Dialogues in Human Geography*, **3** (3) 274–279.

Bermingham, A. and Smeaton, A. (2011) On using Twitter to monitor political sentiment and predict election results. In: *Proceedings of the Workshop on Sentiment Analysis Where AI Meets Psychology (SAAIP 2011)* Chang Mai, Thailand (2–10).

Best, R. and Meikle, J. (eds.) (2015) *Measuring Construction: Prices, Output and Productivity* (Abingdon: Routledge).

Bloom, Burton H. (1970) Space/time trade-offs in hash coding with allowable errors. *Communications of the ACM*, **13** (7), 422–426.

Borgman, C. L. (2016) *Big Data, Little Data, No Data: Scholarship in the Networked World* (Cambridge, MA: The MIT Press).

Boton, C., Halin, G., Kubicki, S. and Forgues, D. (2015) Challenges of big data in the age of building information modeling: a high-level conceptual pipeline. In: Luo, Y., (ed.) *Proceedings of the 12th International Conference on Co-operative Design, Visualization and Engineering*, Mallorca, Spain, September 20–23 (48–56).

Breiman, L., Friedman, J. H., Olshen, R. A. and Stone, C. J. (1984) *Classification and Regression Trees* (Monterey, CA: Wadsworth, Inc.).

Capelhorn, P. (2012) *Whole Life Costing: A New Approach* (Abingdon: Routledge).

Cave, A. (2017) What will we do when the world's data hits 163 zettabytes in 2025? *Forbes*, 13 April. www.forbes.com/sites/andrewcave/2017/04/13/what-will-we-do-when-the-worlds-data-hits-163-zettabytes-in-2025/.

Coelho, V. N., Coelho, I. M., Coelho, B. N., Reis, A. J., Enayatifar, R., Souza, M. J. and Guimarães, F. G. (2016) A self-adaptive evolutionary fuzzy model for load forecasting problems on smart grid environment. *Applied Energy*, **169**, 567–584.

Coffman, D. (2015) The political economy of corn markets. In: Baranzini, M. L., Rotondi, C. and Scazzieri, R. (eds.) *Resources, Production and Structural Dynamics* (Cambridge: Cambridge University Press).

Eastman, C., Teicholz, P., Sacks, R. and Liston, K. (2011) *BIM Handbook: A Guide to Building Information Modeling for Owners, Managers, Designers, Engineers and Contractors*, 2nd ed. (Hoboken, NJ: John Wiley & Sons).

Egbu, C. and Robinson, H. S. (2005) Construction as a knowledge-based industry. In: Anumba, C., Egbu, C. and Carillo, P. (eds.) *Knowledge Management in Construction* (Oxford: Blackwell Publishing Ltd.).

Emmitt, S. and Gorse, C. (2003) *Construction Communication* (Oxford: Blackwell Publishing Ltd.).

Gandomi, A. and Haider, M. (2015) Beyond the hype: Big data concepts, methods, and analytics. *International Journal of Information Management*, **35** (2), 137–144.

Gordon, A. D. (1996) Hierarchical clustering. In: Arabie, P., Hubert, L. J. and De Soete, G. (eds.) *Clustering and Classification* (River Edge, NJ: World Scientific).

Granger, C. W. J. (1969) Investigating causal relations by econometric models and cross-spectral methods. *Econometrica*, **37** (3), 424–438.

Granger, C. W. J. and Newbold, P. (1974) Spurious regressions in econometrics. *Journal of Econometrics*, **2**, 111–130.

Grenadier, S. R. (1995) The persistence of real estate cycles. *Journal of Real Estate Finance and Economics*, **10**, 95–119.

Jaffar, N., Abdul Tharim, A. H. and Shuib, M. N. (2011) Factors of conflict in construction industry: A literature review. *Procedia Engineering*, **20**, 193–202.

Jensen, A. and Varano, M. (2011) Technical due diligence: Study of building evaluation practice. *Journal of Performance of Constructed Facilities*, **25** (3) 217–222.

Kennedy, P. (1998) *A Guide to Econometrics*, 4th ed. (Oxford: Blackwell Publishing Ltd.).

Kitchin, R. (2013) Big data and human geography: Opportunities, challenges and risks. *Dialogues in Human Geography*, **3** (3), 262–267.

Kitchin, R. (2014a) The real-time city? Big data and smart urbanism. *GeoJournal*, **79** (1), 1–14.

Kitchin, R. (2014b) Big data, new epistemologies and paradigm shifts. *Big Data & Society*, April-June, 1–12. doi:10.1177/2053951714528481.

Kshetri, N. (2014) Big data's impact on privacy, security and consumer welfare. *Telecommunications Policy*, **38** (11), 1134–1145.

Laney, D. (2001) 3D data management: Controlling data volume, velocity and variety. *META Group Research Note*, **6**, 949. https://blogs.gartner.com/doug-laney/files/2012/01/ad949-3D-Data-Management-Controlling-Data-Volume-Velocity-and-Variety.pdf.

Langston, C. (2015) Performance measures for construction. In: Best, R. and Meikle, J. (eds.) *Measuring Construction: Prices, Output and Productivity* (Abingdon: Routledge).

Liu, Y. and Xie, T. (2018) Machine learning versus econometrics: Prediction of box office. *Applied Economics Letters*, 1–7. https://doi.org/10.1080/13504851.2018.1441499.

Manyika, J. and Chui, M. (2011) *Big Data: The Next Frontier for Innovation, Competition, and Productivity* (Washington: McKinsey Global Institute).

McClelland, J. L., Rumelhart, D. E. and the PDP Research Group. (1986) *Parallel Distributed Processing: Explorations in the Microstructure of Cognition. Volume 2: Psychological and Biological Models* (Cambridge, MA: The MIT Press).

Olanrewaju, A. L., Tan, S. W. and Kwan, L. F. (2017) Roles of communication on performance of the construction sector. *Procedia Engineering*, **196**, 763–770.

Pärn, E. A., Edwards, D. J. and Sing, M. C. P. (2017) The building information modelling trajectory in facilities management: A review. *Automation in Construction*, **75**, 45–55.

Provost, F. and Fawcett, T. (2013) Data science and its relationship to big data and data-driven decision-making. *Big Data*, **1** 51–59.

Rumelhart, D. E., McClelland, J. L. and the PDP Research Group. (1986) *Parallel Distributed Processing: Explorations in the Microstructure of Cognition. Volume 1: Foundations* (Cambridge, MA: The MIT Press).

Russell, S. and Norvig, P. (2002) *Artificial Intelligence: A Modern Approach*, 2nd ed. (Upper Saddle River, NJ: Prentice-Hall).

Saleuddin, R. and Coffman, D. (2018) Can inflation expectations be measured using commodity futures prices? *Structural Change and Economic Dynamics*, **45**, 37–48.

Tapscott, D. and Tapscott, A. (2016) *The Blockchain Revolution: How the Technology Behind Bitcoin Is Changing Money, Business and the World* (New York, NY: Portfolio, Penguin Books).

Thom, R. (1972) *Stabilité Structurelle et Morphogénèse* (Reading, MA: W.A. Benjamin).

Thom, R. (1991) *Prédire N'est Pas Expliquer* (Paris: Eshel).

Van De Wetering, J., Dixon, T. and Sexton, M. (2016) *Smart Cities, Big Data and the Built Environment: What's Required?* European Real Estate Society (ERES). https://Econ Papers.repec.org/RePEc:arz:wpaper:eres2016_336.

Wall, M. (2014) Big Data: Are you ready for blast-off? *BBC News*, 4 March. www.bbc.com/news/business-26383058.

Wei, Y. and Miraglia, S. (2017) Organizational culture and knowledge transfer in project-based organizations: Theoretical insights from a Chinese construction firm *International Journal of Project Management*, **35** (4), 571–585.

Yu, M. and Ive, G. (2008) The compilation methods of building price indices in Britain: A critical review. *Construction Management and Economics*, **26** (7), 693–705.

Zeeman, E. C. (1976) Catastrophe theory. *Scientific American*, **234**, 65–83.

Editorial comment

In this, the final chapter, the authors put forward some ideas regarding the future of the construction industry. They look towards the disruptions that are changing business models in many industries (e.g. Uber, Airbnb and Amazon) and speculate on how similar disruptions may affect construction.

Central to their argument is the notion that construction firms must lift their game if they are to remain competitive in an increasingly global market. They point to the need for fundamental change if the industry is to avoid being overtaken by disruptive forces that could seriously damage their chances of survival. They suggest that a part of any industry response to disruption must lie in finding reliable metrics that can be used to gauge industry improvement. They offer an outline of a method for assessing performance improvement based on project data that they claim can be easily recorded and can be aggregated to provide measures of improvement at both the firm level and the national level.

They intentionally depart from the conventional measures such as productivity, project cost and completion within budgets and/or schedules. Many of the difficulties associated with such metrics are discussed in Chapter 1. A significant part of the rationale for the choice of the metrics proposed here is that there has been a conscious attempt made to find alternatives to the usual metrics precisely because of the problems that arise when, for example, we try to assess productivity at any level above on-site work items (e.g. activities such as laying bricks where obvious measures such as number of units laid per day per person are relatively easy to record) or we are faced with project costs expressed in different currencies or recorded at different times.

The measures proposed are all relative, not absolute, measures and thus the emphasis is entirely on comparisons. The intent is that improvement targets can be set and then progress towards those targets can be monitored using available data to compare outcomes over time. This approach may be best suited, in the first instance, to smaller, more nimble organisations that are, by their nature, better placed to lead the way. Ideally such monitoring would be done on a voluntary basis, but if there is value in the practice there is nothing to stop governments making participation mandatory, perhaps as part of wider strategies aimed at improving industry performance in areas such as productivity and safety. Such improvements are generally considered to be both necessary and desirable.

It can be argued that productivity in construction has not improved in recent years as might have been expected, given the wide-ranging and rapid developments in technology that have occurred; this is part of the authors' reason for seeking alternative ways to assess changes in productivity. Similarly, while there has been a steady decrease in serious and fatal accidents on sites in recent years in some places (Australia and the UK, for example; see Safe Work Australia 2015; Glenigan 2015), such incidents do still occur and until the industry can report that it is, if not accident-free, at least fatality-free, there will always be room for improvement.

The focus on comparisons presents its own problems, as valid comparisons rely on like being compared with like. Building projects are at best similar but seldom, if ever, identical, and they are often fundamentally and profoundly different. Some firms may build nothing but houses or petrol (gas) stations but for every one of those there are many that construct a wide variety of building types: warehouses, shopping centres, libraries, apartments and so on. Data relating to, for example, time to construct will vary according to, *inter alia*, project type, size, quality and complexity. Such data requires some sort of normalisation if credible comparisons are to be made. The authors have made some attempt to do this based on floor area of a variety of projects but there is room for further development of this aspect of the framework. The floor area approach may be suitable for general building, but such a method would be unlikely to work for infrastructure/ civil engineering projects or even highly specialised buildings such as high-tech laboratories and factories. Once again, the intent here is to introduce an alternative approach to a measurement problem rather than to present a mature, well-tested process. Equally, the authors do not suggest how targets may be achieved; the focus is purely on how progress might be assessed, by measurement and comparison of current performance with prior performance.

The proposed approach, therefore, should be viewed as a work in progress. It should also be noted that while the chapter contains some references to the UK and Australia the intent is that the proposed approach can be applied globally; references to specific countries are for illustration only.

12 A proposed framework for measuring future construction industry performance

David Chandler, Mary Hardie,
Srinath Perera and Craig Langston

Introduction

Measuring construction performance improvement is the new frontier in observing transformations that will, in the future, redefine the global construction industry. Establishing industry-wide performance measures that provide a line of sight for governments, industry stakeholders and clients to effectively monitor and interpret the many claims made about achievements in this area prove elusive. Despite little disputation over the potential for construction performance to be improved, the necessary connections to make this happen have not yet been able to fully align (Holt and Goulding 2016).

There are three main forces threatening to reshape today's construction industry. These explain why new performance measures are now important and necessary. They are:

1. construction fully engaged with the digital economy;
2. construction rapidly becoming industrialised; and
3. construction being organised and procured in a more dynamic global marketplace.

Construction in the near future will be characterised by construction firms that embrace the above forces and adapt accordingly. In this context, modern constructors need tools to demonstrate they are measurably better than their traditional construction counterparts. This will lead to tangible benefits for construction's future customers.

The aim of this chapter is to conceptualise a new framework for measuring future construction industry performance. Traditional price measures, such as construction cost per square metre, suffer from difficulties concerning time, location and specification that make comparison within and across national borders difficult. New paradigms for construction, including greater focus on off-site processes, are game changers in accounting for industry transformation. Non-price measures offer the possibility of benchmarking performance more openly and without the impediment of price conversion. The focus is on construction firms and the delivery of projects that meet the expectations of society in general and customers in particular.

We first look at the need for change and then outline a vision of construction in the future. Finally, a conceptual framework is presented using a small number of key non-price measures that are thought to be critical to an industry that is undergoing transformation. These measures are the number of days of on-site activity (start to finish), the volume of construction waste sent to landfill, the average number of workers on-site per day and lost time due to injuries, assessed in terms of reductions over time. Further measures to include the after-build (operational) phase can also be added. The framework is discussed in terms of its implementation strategy before final conclusions are drawn.

Modern construction is defined as methods of procurement that lead to performance improvement compared to traditional building processes. Ultimately these methods are intended to deliver higher product scope for less cost, in less time and at less risk. Put another way, modern construction participants must work smarter, better, safer and faster to deliver projects that represent higher levels of value, efficiency, speed and innovation, with lower levels of complication and impact (Langston 2013). Doing this consistently, project after project, year after year, on a global scale, is the industry's grand challenge.

The need for change

There are many new technologies and construction methodologies all pressing to make their case for how they are leading the charge to improve construction certainty and performance. These include building information modelling (BIM), design for manufacture and assembly (DfMA), lean construction, big data, off-site fabrication, constantly evolving applications that enable smarter business-to-business dealings, monitoring of safety, documentation transmittal and storage, quality and certification management, and a plethora of cloud-based software applications that enable remote project, enterprise and built asset management. Despite all of these seemingly efficiency-based developments, the costs of construction continue to rise and much-published data points to industry performance having plateaued (Agarwal *et al.* 2016).

While the measurable advantages of a modern construction industry should be showing up, they are not. The traditional construction industry has remained steadfastly measurement resistant and so have its practices.

Indeed, these questions have been the subject of frustration for initiatives such as the industry performance improvements aspired to in Sir John Egan's report on *Rethinking Construction* (Egan 2002) and the UK's *Construction 2025* industrial strategy (HM Government 2013). There are similar less directed endeavours in the United States, Australia, New Zealand and Malaysia. The weakness in all of the industry improvement strategies worldwide is the industry's fragmentation across its traditional parts, the professions and jurisdictional differences mostly influenced by self-interest and the often claimed need for 'others' to change their behaviours first. Larger organisations have sought to hold on to their traditional market advantage, which is determined by scale and influence. Smaller organisations have been constrained by industry governance or representations that enshrine an unsophisticated 'one size fits all' approach to supply chain engagement and industry policy development (i.e. over-simplification of best practice).

Other barriers to construction industry performance measurement have been the obvious differences between project type, complexity, scale, industry capacity and location. Another is sheltering of perceived commercial competitive advantage or intellectual property to assert a market position for the next deal. There is always a reason to escape being held to account to deliver the measured industry performance step change that always seems within grasp but somehow escapes for another day. The reality is that the traditional defences available to those who have taken solace in the status quo are crumbling. Miller *et al.* (2009) cited a momentous shift challenging old industrial models that have served construction well for centuries but now face calamitous post-industrial stresses. They concluded that these are becoming increasingly unworkable in our networked world. Miller *et al.* (2009) do not leave a lot to the imagination for those who want to continue resisting change and/or fail to see the future through a customer lens. They claim traditional construction models are losing relevance.

The trend of large technology firms disrupting traditional business models and monopolies will be ignored by slow-moving construction enterprises at their peril. Today, new collaborations are occurring amongst suppliers and clients. They are often driven by dissatisfaction. The digital economy will coalesce with moves in the collaborative economy. Companies such as Uber and Airbnb are often cited for the disruptions they are causing to their respective industries. Overlooked are the impacts emerging in the area of dispute resolution and the regulation of behaviour out of these disruptions. These disruptions will in time flow into construction, just like the impacts that are happening from the evolving concepts around mobility services. A relevant example is how the worlds' insurers are being forced to rethink new mobility service risks and how they price them. Consider then, the progressive insurance impact from the components of construction being given digital fingerprints that will soon follow them through manufacture to installation and throughout their physical life cycle.

With these trends in mind, the importance that will attach to making construction more innovative and measurably competitive is significant. Understanding the core attributes that affect construction performance will be fundamental in the new construction world. Energising local construction to be more competitive will require access to high quality pre-competitive (currently confidential) data that has tracked performance over time. It will be this data that enables strategically driven innovation investment to be effectively informed and directed. The data must be simple to gather, transparent, without bias and widely available. Investigating which industry and enterprise initiatives might offer the best returns depends on the quality of these insights.

A vision for future construction

Many industries have experienced the disruptive impact of change generated by a combination of new technology and globalised markets (Christensen *et al.* 2002).

The construction industry in countries like Australia, for example, has often been seen as immune to this disruption because its projects are delivered locally in a specific national and geographic context. This situation is changing because of international competition and the adoption of new technologies. The growth in DfMA, use of robots, augmented reality, off-site manufacture, volumetric, panelised and sub-assembly construction will increasingly mean that local markets face global competition. As Best (2010) found, value for money comparisons (price measures) are complex and difficult to make across different construction projects and different jurisdictions. This is likely to be more problematic in the global marketplace unless standards are developed about how to judge project performance.

A great deal of research effort has gone into understanding the different ways to measure construction (e.g. Kenley 2014; Yi and Chan 2014). Because of the complexity of measurement factors involved, little existing research has concentrated on measuring construction performance across a spread of project types. The tendency is to focus on comparisons of a particular type of construction across several countries (Chan 2011; Abdel-Wahab and Vogl 2011; Langston 2014; Nasir *et al.* 2014) or on a single class of building in one country (Chan 2009; Guerrini 2013; Newton *et al.* 2012). Data reliability is paramount (Rojas and Aramvareekul 2003). It is therefore crucial that a research focus be developed on achieving common definitions, measurement indices and benchmarking protocols in order to guide reporting of construction trends (Crawford and Vogl 2006). The establishment of new baseline measures would enable targeted improvement in terms of overall project costs, time from inception to completion, constructed quality and compliance, embodied energy, waste reduction, greenhouse gases, construction safety and projected future maintenance costs. These would also enhance the prospects for higher quality and lower performance risk.

Previous research by Dubois and Gadde (2002) suggested that the construction industry is characterised by (1) particular complexity factors owing to industry specific uncertainties and interdependences and (2) inefficiency of operations. Their study analysed the operations and behaviour of firms as a means of dealing with complexity. They found that the industry as a whole is a loosely coupled system, and taking this as a starting point, the couplings among activities, resources and actors were analysed in different dimensions. The pattern of couplings built on two interdependent layers: tight couplings in individual projects and loose couplings based on collective adaptations in the permanent network. It was concluded that the pattern of couplings seems to favour short-term productivity goals while hampering innovation and learning in the longer term.

Ofori (2001: 41) found that an important reason for the lack of progress in construction industry development is '*the absence of measurable targets*'. He proposed a raft of indicators, including construction output, export volume, volume of material imports, employment, human resource development, affordability, cost, safety and sustainability. Alternatively, Willis and Rankin (2012) argued that construction performance could be assessed from industry maturity modelling based on a set of thresholds that must be passed in order to achieve sustainable improvement and using after-the-fact measures from a statistical sample of construction projects.

Crawford and Vogl (2006) provided an overview of methods used to measure productivity in the construction industry. The advantages and disadvantages of average labour productivity and total factor productivity measures were discussed in detail, and the relationship between these two measures was established both theoretically and in an application at the industry level. The usefulness of any productivity measurement framework for policymakers and industry practitioners alike depends crucially on the extent to which it enables the identification of the underlying drivers of productivity. This requirement necessitates an approach that involves formally describing the production process and explaining as much as possible of construction output in terms of the quantity and quality of inputs that are used to generate it. Whilst it is accepted that data requirements are a major constraint to such an approach, it is suggested that by establishing a robust measurement framework, data deficiencies can be defined more easily. Guidance on areas where improvements are needed is provided and the focus of future research should be on creating new datasets and improving those that already exist.

Doloi (2008) used a structured questionnaire survey to analyse labour productivity in construction projects in Melbourne, Australia. The survey was formulated with 72 questions covering three broad categories, namely, project planning, incentives/disincentives and job satisfaction. Perceived best practices that impact on the improvement of productivity on-site were identified from 19 targeted experts. Analytical hierarchy process showed that the biggest influence on productivity was planning and programming. Indeed, Jarkas (2015) later suggested that buildability issues partly explain low productivity in the construction industry.

Chancellor and Abbott (2015) looked at the effect of shadow economic activity on productivity growth in Australia and found it to be significant due to use of cash-in-hand and barter trade. Construction arguably has one of the largest shadow economies of any Australian industry sector, and it continues to grow. The shadow economy is harmful to taxation revenue, business competition, safety and quality standards. They argued that it is important to capture shadow activity as part of any approach to construction performance measurement. Chancellor (2015) also found that expenditure on wages and research and development were major factors to drive growth at the national level. Unmeasured construction productivity hinders the arrest of this economic leakage.

Under-measurement of the sector's impact on the economy elevates the importance of lifting construction performance; see de Valence (2001) on satellite accounts for the built environment sector, Ashworth and Perera (2015) on the economic significance of construction and Foulkes and Ruddock (2007) on defining the scope of the construction sector. All point to under-measurement. Measuring the future impact of construction in the economy may become increasingly difficult as more construction fabrication is moved off-site into a manufacturing setting. The trends indicate that construction on-site will progressively be assembly orientated. Tracking future construction employment and off-site construction-related enterprises who service the sector will require further consideration in national accounting.

More recently, Vereen *et al.* (2016) presented a metric for quantifying productivity using labour and cost data from a sample of 100 typical construction activities. They found that the quality of work over time was difficult to capture as data that can be fed into consistent and repeatable metrics is generally neither collected nor kept by the construction industry. Vogl and Abdel-Wahab (2015) went further to advocate the establishment of an 'international benchmarking club' to inform policy on performance improvement.

In a globalising industry, construction businesses will need to become more agile and adaptive as they face competition from client-focused and information technology-savvy entrepreneurs changing the way projects are delivered (Sezer and Bröchner 2014). The establishment of benchmarking metrics for non-price performance measures would inform clients of alternative project delivery options in the transforming construction environment that lies ahead. New managerial, contractual and compliance initiatives will also be needed to respond to the impact of technological change and to lift industry effectiveness (Salmon 2012). New change drivers now seem to be aligning, which will be conducive to any new benchmarking framework.

Gambatese and Hallowell (2011) argued that creative innovation was vital for successful and long-term company performance in the construction industry. They found that understanding the innovation process, how it can be enriched and how it can be quantified were key elements to managing innovation and creativity. Case studies of construction projects in the United States revealed that idea generation, opportunity and diffusion were crucial (Hardie 2016). Applying these practices also leads to improved levels of innovation via better communication among project team members, integration of design and construction processes, greater efficiency and development of unique ways of learning. This can be demonstrated to apply for all sizes of companies and projects.

In recent years, several established industry sectors have experienced the impact of what Christensen *et al.* (2002) termed 'disruptive innovation'. Disruption is said to have occurred when established businesses are out-competed and displaced by newly established ones that introduce new technologies or processes. Using grounded theory and a complete census of the case histories of every model of disk drive developed between 1976 and 1992, Christensen (1992) derived a theory of disruption that presented two intersecting trajectories of performance improvement. He noted that in the early stages of the introduction of technological breakthroughs (disruptive technologies), new entrant enterprises nearly always win. This is because the functionality of the current offerings is not good enough to meet their customers' needs. However, incumbent or established enterprises nearly always win when the technology reaches the point of being a sustaining innovation or one that meets or surpasses the customers' requirements (Christensen *et al.* 2002).

Examples of disruption have been documented in industries as diverse as transportation (Cho and Parsons 2015), photography (Čiutienė and Thattakath 2014), newspapers (Karimi and Walter 2015), retailing (Christensen and Tedlow 2000), recorded music (Moreau 2013) and computer graphics (Pimentel 2010). In addition, research by Jang (2013) has identified seven disruptive innovations that are likely to have an impact in the future. These include wearable computers, 3D printing, context awareness technology, driverless cars, ultra-light materials, gene therapy, and next-generation batteries.

More recently, a study of disruption in construction by New Zealand's BRANZ (Clark-Reynolds and Pelosi 2016) found seven types of construction change drivers. These included mass customisation, products becoming services, services becoming products, separation of ownership and use, the sharing economy, agile design and construction, and agile planning. The study identified the constraints holding the transformation of construction back as uninformed clients, traditional industry, regulation, information, contractual constraints and banking finance.

Despite the predicted widespread nature of disruptive innovation in the 21st century, the construction industry has mostly avoided large-scale disruption because, in spite of the globalisation in mega-projects, most construction is locally based and delivered within a specific national context and regulatory system. Some research suggests that this may well be about to change. Examples of potentially disruptive innovations that have the capacity to generate significant change in the construction industry include a large range of off-site construction systems, robotic construction, 3D printed building components and systematised modular construction elements (Gibb and Isack 2003; Blismas *et al.* 2006; Tam *et al.* 2007; Yu *et al.* 2013; Martínez *et al.* 2013; Landers 2015).

Perceived inefficiencies in the industry, along with widespread dissatisfaction with the quality of construction delivery, leaves the industry vulnerable to disruptive competition from start-up enterprises that harness value by using a combination of digital technology and new production techniques. Research that looks into ways to measure construction performance across national contexts and varied project types, which could accelerate this change process, is currently being pursued (Langston 2014; Chan and Hiap 2012). Significant drivers of change are acting on the industry and are likely to have an accelerating transformational impact.

Proposed conceptual framework

At the heart of a modern construction industry must be a new psyche addressing how a measurably better deal will be delivered to its customers. This means more for less. It does not mean that those who work most effectively as a part of construction's future value chain should work for less, but they should work smarter, better, safer and faster. In a digital, industrialised and global construction economy, there is no room for selective gestures to any of these core drivers. They must all contribute to a measurably better deal, just as they have in other modernised industries. It does not matter how far industries such as banking, insurance, electronics, mobility services, education, health or government service delivery may have come already, all realise that unless they continue to raise their offering to customers, it is game over. Construction is now part of this mix. This will result in new collaborative workarounds that disrupt construction's traditional constraints. It is a global issue that is not confined to specific countries.

There are many sources of insight into the pathways that can be taken to pursue industry performance improvement. Motorola's *Six Sigma* program, for example, demonstrates a set of techniques that collectively pursue a defined sequence of steps that have specific value targets, such as reducing process cycle time, reducing pollution, reducing costs, increasing customer satisfaction and increasing profits. *Six Sigma* is directed at improving the quality of output of a process using statistical methods, supported by people within the organisation who are expert in these methods (Cherrafi *et al.* 2016). The same should apply to modern client-focused construction enterprises.

The UK's *Construction 2025* (HM Government 2013) has adopted a number of key performance goals to be attained by 2025 (compared to 2013 levels). These include:

- iowering costs by 33%;
- improving delivery speed by 50%;
- iowering construction delivery and after-build greenhouse gas emissions by 50%;
- improving exports by 50% through reducing the trade gap between total exports and total imports for construction products, materials and services.

These are worthy aspirational benchmarks, and few in the construction industry today would argue that they are unachievable. But most will then make a rejoinder that they are not in control of all of the processes that will deliver these benefits. The difficulty to date has been that the way forward is viewed through the traditional 'price-measure' lens. A more viable means of pursuing these performance improvements will more likely come from the use of 'non-price' measures. Once these measures and their benefits are captured, it will then be possible to develop a set of simplified indicators that point to performance improvements occurring in projects over time.

There are three critical phases in the construction performance cycle that have historically been the major points of process disconnect:

- the pre-construction phase including project definition, planning and procurement;
- the construction delivery phase including fabrication, assembly and acceptance;
- the after-build phase including operations, maintenance and sustainability.

In other industries, these disconnects have dissipated as each has become part of a seamless flow of processes that lead to satisfying the final customer experience described by McKinsey (2016) and evidenced by global companies such as Motorola and General Electric. These developments will require re-calibration of the industry's customer value proposition, its enterprises (value chains) and their projects (products).

In traditional construction, a growing barrier to achieving seamless outcomes is risk aversion. In this context the providers of the parts of construction look to defences that will shield them from exposure in the event that the sum of the parts under-delivers on contracted performance standards, fitness for purpose or functional durability. The construction industry deals with its habitual defences in the case of accountability by placing strict caveats on the extent of liability and the effects that acts or omissions by others may have. Another defence is tendering late stage alternates that are to be assumed by the client as suitable. These combine to affect work flow effectiveness, falling productivity, variable quality and the industry's rising costs.

On-site, constructors are hardly able to detect many of the claims made by their off-site counterparts, and the customers of construction will not reap the sum of benefits claimed. One can argue that construction's customers in the end deserve what they get because they sit at the front of procurement culture that has enshrined practices that are the root cause. So, too, have their traditional consultants and advisors.

The reality is that other industries, like retailing, telecommunication and electronics, have worked around these perceived impediments to take on single point accountability for 'in-full and on-time' delivery. They have done this because they recognise the growing dissatisfaction amongst their customers with the old status quo and where, in a global industrialised and digitised marketplace, the cost and resources required to deal with system failures across distance, differing jurisdictions and consumer legislation have made best practice a more viable option than the minimal alternative. Construction is now at that crossroad.

Modern construction is a term that describes the new era that is now simultaneously reshaping the industry's organisations, projects and people. This era will require construction policy and legislation to be reshaped. These changes will have profound implications for the skilling of future constructors and subsequently defining and powering their enterprises and projects. Perhaps the most significant difference between traditional construction and modern construction is that future performance will be more measurable and transparent. First movers will realise that they no longer need to be inhibited by the performances of the less innovative. When valuable competitive advantage can be leveraged by their ability to demonstrate new performance trajectories, they will soon make the industry's unproductive laggards obsolete.

While performance benchmarks may have been avoidable once, they seem inescapable from here on. Motorola found that the simplest of measures worked best. These were the ones for which data was the easiest to collect, that were simple and the easiest to interpret. Their interpretation was compelling and their selection had system-wide effects. Lean construction (LCI 2017) deploys many of these techniques and issues escalation protocols. The difficulty for lean construction is that it has limited, if any, industry accepted measures or benchmarks against which performance can be compared. Most deploy the performance of the last project as the benchmark for the next. Where they exist, these measures are often personal to the customer, program or project. Alternatively, the metrics are held by a consultant or contractor as intellectual property to be applied to secure the next engagement. In the modern construction era, measurement transparency must prevail.

The literature speaks of data simplicity and collection targeted to drive maximum benefit. Pre-construction effectiveness (PCE) and construction delivery effectiveness (CDE) are proposed measures that adopt this principle; they are discussed further in Table 12.1. The non-price datasets for time, workforce, waste and safety are recorded as a contractual requirement or in the normal administration of projects. The PCE incorporates time and waste measures as these are most able to be influenced prior to construction commencing. The CDE incorporates workforce and safety measures as these are most able to be influenced during construction. Each measure is considered to have equal weight in the proposed framework. Optimal effectiveness will occur when all measures are progressed concurrently. The suggested measure targets are based on extensive consultation with industry players in the UK, Europe and Australia, and are consistent with those aspirations in HM Government (2013).

There is no better place to observe, and thus to verify the effectiveness of, pre-construction and actual construction management than on-site where projects are performed. It is here that the utilisation of time, on-site infrastructure, supervision, workforce inputs, materials, waste, disruption and duplication can be physically measured. It is on-site that the traditional packaging, sequencing and skilling of the workforce play out most visibly. It is on-site where all prior assumptions about constructability (buildability) and planning are laid bare.

Collection of the data needed for measures such as those in the framework is currently required in many Western countries, either contractually or statutorily. The point for now is the importance of starting with available, simple and easy to interpret data until this approach becomes routine. Holding back for more complex measures plays into further obfuscation and resistance. A simple source of performance data will also point to opportunities for new construction components, assembly methods and services to be developed that target specific improvements and potentially point to new enterprise opportunities. These should help in identifying new local pre-build and on-site innovations that feed into local economies by contributing to domestic employment and investment.

The subsequent after-build phase (ABE) measures may be developed following an initial period of collecting and considering the base data (i.e. the data required for the PCE and CDE measures) described in this chapter. Keeping the PCE and the CDE measures as simple as possible is paramount. All the research done around total quality management and *Six Sigma* indicates that only a few measures are needed to influence the majority of the outcome. The Pareto effect (sometimes known as the 80:20 rule) applies. Refinement can occur over time. As more studies are conducted the datasets to support the wider use of non-price measures will become apparent.

Each measure can be assessed in terms of reductions over time. It is suggested that performance improvement needs to be assessed within a scale of five years

Table 12.1 Proposed conceptual framework

Measure	Reason	Application
Measure 1 – PCE input (Time) Records the total calendar days of elapsed time from the moment work starts on-site until final client acceptance is achieved. Measuring calendar days allows for multi-jurisdiction consistency and acknowledges the time cost of finance.	To track progress towards reducing on-site construction durations by 25% within three years (50% in five years). The measure provides industry with the benchmarks necessary to drive new project assembly methodologies and pre-build innovation. The goal should be to bring the pre-build manufacture relationship into the room during project definition based on predetermined cost and performance targets.	This measure puts focus on those scoping, specifying, pricing and planning projects to re-imagine how their inputs can improve on-site performance and duration. It should enable a rethink of the potential to share and lock in project definition, while looking to incorporate more standardised assembly methods and practices. There should be no lessening of the value or interparty performance accountability imperative.
Measure 2 – PCE input (Waste) Recording the amount of physical waste (m^2) removed from the project (excluding excavation, site remediation and the demolition of pre-existing structures) commencing at the date of on-site construction starting through to completion.	The objective targets a 50% reduction of construction waste within three years (80% in five years). There is considerable opportunity to minimise construction waste even as current on-site fabrications of traditional construction are progressively replaced by the assembly of more off-site or pre-build manufactures. Feedback to the pre-construction and off-site fabrication teams will enable early and more visible correlation between the BIM, DfMA and project scheduling technologies.	The earlier planning to minimise construction waste commences and a line of direct origin is drawn, the sooner smarter specification and work packaging can be applied to future construction. This measure should draw the pre-construction team to the construction workface to investigate what works and what would work better. Minimising time and waste will drive overall construction system improvement and effectiveness.

(Continued)

Table 12.1 (Continued)

Measure	Reason	Application
Measure 3- CDE input (Effort) Recording the total workforce inputs (average number of workers/day) including all site-based management and supervision staff from the moment work starts on-site until final client acceptance is achieved. Any paid time lost during inclement weather, industrial delays or scope change should be included.	Measuring the effective utilisation of on-site workforce inputs is aimed at achieving a reduction of 30% over the initial measurement period of three years (50% in five years). This measure should enable a line of sight to increasing incorporation of off-site manufactured components. This measure should enable a shift from nominally >30 trade to <10 advanced assembly based self-managing and performing sequential packages.	Challenges traditional approaches to project work packaging, procurement methods, data sharing, self-management, progress tracking, work acceptance, compliance and payment recognition. This measure should also encourage a differentiation to be made about the best ways to deploy mass fabrication while making room for mass customisation during the initial build and over the life cycle.
Measure 4 – CDE input (Safety) Recording the accident lost time injury frequency (LTI) rate (including rehabilitation duration) arising from all work performed on-site.	A 50% reduction in existing experience rates should be achieved within three years (80% in five years). There should be no compromise of safety in deploying a new industry culture based on the notion of 'better, smarter, faster, cheaper'. This measure should shift the current focus on safety and industrial disruption to one of smarter work, new skills and capability recognition.	There is no more important issue to be managed by the lead contractor and no one better placed to see that high quality pre-construction methods planning, on-site implementation protocols and integration of on-site performance technologies across the assembly team occurs. Off-site fabrication, for example, will have a strong correlation with lifting construction productivity and improving safety.

(i.e. 'then') from implementation of data collection (i.e. 'now'). Figure 12.1 illustrates the generic approach for evaluation of progress in transformation.

It is expected that reductions in the PCE and CDE (as well as future ABE) non-price measures will deliver cost savings rather than cost burdens. The objective is to deliver simultaneous reductions in all proposed measures. The index for PCE is computed as the mean percent reduction in the duration of the on-site process and the volume of waste sent to landfill, between 'now' and 'then'. The index for CDE is computed as the mean percent reduction in the average number of workers on-site and lost time due to injuries, over the same time period.

To demonstrate how this approach could work in practice, an example might help. Let's assume in 2014 a company built four projects and total construction time was 700 days, generated a total of 2,000 m³ of waste, an average of 22 workers per day were on-site, and a total of 200 hours was lost due to injuries. The four projects were two small shopping centres, a petrol station and a community centre having a total value of $30 million and a combined floor area of 11,000 m². In 2018 the company built five projects: a block of apartments, a small shopping centre, two factory/warehouses and a surf club, having a total value of $45 million and a combined floor area of 15,000 m². Total construction time in this case was 700 days, a total of 3,000 m³ of waste was generated, an average of 25 workers per day were on-site, and a total of 200 hours was lost due to injuries. In terms of PCE and CDE, did the company improve, go backwards or deliver similar performance between 2014 and 2018?

One approach to answer this question involves using non-price measures and normalising the performance according to project size. Table 12.2 summarises the performance achievements between 2014 and 2018.

Graphs can be created for each performance measure showing 2014 at 100% and 2018 at 73.4%, 110.0%, 85.0% and 73.4%, respectively. Overall, PCE (time and waste) is 8.3% better and CDE (effort and safety) is 20.8% better. Combining PCE and CDE suggests the company has performed 14.55% better over five years.

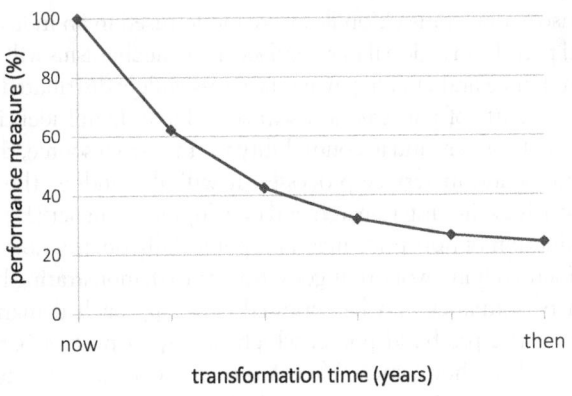

Figure 12.1 Generic evaluation model

Table 12.2 Evaluation example

Measure	2014		2018		Performance
Time (faster)	700 days/11,000 m²	= 0.0636	700 days/15,000 m²	= 0.0467	26.6% better
Waste (better)	2,000 m³/11,000 m²	= 0.1818	3,000 m³/15,000 m²	= 0.2000	10.0% worse
Effort (smarter)	22 people/11,000 m²	= 0.0020	25 people/15,000 m²	= 0.0017	15.0% better
Safety (safer)	200 hours/11,000 m²	= 0.0181	200 hours/15,000 m²	= 0.0133	26.6% better

It is concluded that waste minimisation required greater focus by the company. If monitoring had taken place annually, then evidence of progressive achievement of performance targets during transformation would have been available. The above approach could apply to a company, a geographical area or an entire industry/country.

Discussion

The challenge for the traditional construction industry has been to measurably lift its performance and lower unchecked rising costs. Despite the potential of modern construction technologies and methodologies, achieving these improvements has proved elusive. As the non-price measures become more common and progressively optimised, the question to be answered is, what impact does this have on construction cost and quality? The correlations between time for construction, reducing waste, transforming construction work on-site and substantially improving on-site safety will be unavoidable. It is anticipated that these system improvements will raise quality and lower cost just as has been witnessed in other industries.

When construction transaction flows are modernised in an industry that draws its pieces and parts from a global marketplace, new mechanisms will be required to attest to compliance and enable payments across widely distributed value chains. The pieces and parts of construction will soon have digital identifications that enable a chain of custody and accountability to trace from source, through manufacture, assembly and in-service processes. It will also address the challenges of the drive towards evaluating research and development impact (Holt *et al.* 2016).

The introduction of non-price measures potentially offers a new, easy to adopt and sharply focused framework that goes straight to demonstrating how improved construction performance can be captured. The approach demands that what happens during the pre-build phase, which deploys a host of hard to measure technologies and methods, must have a direct and measurable impact on the physical work to follow. Not having this linkage is unrealistic and dampens the potential to improve construction performance in the future.

Non-price measures are considered to be most influential in setting up the necessary collaboration between pre-build, build and after-build phases in order to deliver better projects and make the most meaningful benchmarked contribution to construction performance improvement and value for money outcomes. An obvious conclusion is that continued downward pressure on construction margins is likely to expose cost burdens sheltered by poor pre-construction inputs, dysfunctional supply chain engagement and risk averse construction practice. Non-price measures will provide a truth of capability and the potential for clients not to subsidise the industry as it makes its transition to the 'modern' era. All of the non-price measures should be weighted equally initially, however, future refinements could include an investigation of possible variations in weightings.

The authors do not promote or recommend one construction delivery method over another. It will be for the more forward-thinking clients and constructors to determine how they will deploy their enterprise technologies, procurement, delivery, chain of custody and performance underwriting initiatives. These will rapidly set the leaders apart from their slower moving counterparts. Continuing to promote 'one size fits all' construction practices is impractical. It is expected that the smaller, more nimble contractors will be the ones who disrupt current practice and push the more established firms out of their comfort zone. There may be projects where these methods could be demonstrated, which would in time enable industry targets to be approached or exceeded.

The industry's transformation will expose new value chain opportunities. Re-organisation of workforce inputs on-site will give rise to new multi-skilled, self-performing and managing assembly businesses. Progressively shifting wasteful and time consuming on-site fabrication to efficient off-site fabrication and pre-assembly environments will deliver better, more cost effective production and traceable quality construction inputs. These transformations will open up new value adding enterprise opportunities and innovative collaborations. Modern constructors will not be sentimental about unproductive customs and practices, nor will they be burdened by their avoidable overheads.

Slow conversion of the targets set in HM Government (2013) seems to have lacked consistent pre-competitive performance measures. The major constructors in the UK seem keen to pursue the potential competitive advantages of new construction technologies as long as they can selectively attest to their time, labour savings and quality achievements. Without a pooled ability to benchmark their collective and individual performances via a more systematic collection of data that consistently quantifies construction performance, the UK's aspirations are likely to take a long time to realise.

The deployment of consistent measures to benchmark construction performances is as valuable to developed economies as it is in developing ones. Developed countries are faced with massive infrastructure expansions and rebuilds. For every avoidable dollar built into projects due to sub-optimal industry performances the trailing economic impact will affect international competitiveness and, eventually, social cohesion. In developing countries, rising labour costs, under-skilled workforces and ever-increasing costs for new construction materials is eroding how much economic and social infrastructure can be delivered through development aid budgets and by their fledgling economies. These conditions affect the potential of those economies becoming viable players in the global marketplace where sophisticated trade of value-added services and building components becomes the product of a digitised and industrialised construction sector.

New construction methods that depend more on on-site assembly of manufactured components methods will call for a rethink of construction workforce capability building. This will involve multi-skilled sequential, advanced work packages being performed along with development of digital communications skills that enable self-directed workforces to directly access real time project information, to self-manage compliance and control quality, safety, problem solving and reporting. The industrialisation of construction's pieces and parts will enable real-time tracking, component verification, in-service performance and energy analysis, project after-care, easier later adaptation and eventual recycling.

The enterprise value creation potential of new pre-competitive data that informs and enables sustainable competitive advantages to be deployed by first mover constructors seems limitless. For constructors who fail to grasp the coming impact of the modern construction era, the future would seem to be comparatively bleak.

Conclusion

There is no industry organisation with an arm's-length mandate to independently collect data that provides an evidence-based platform from which to report on benchmarks that reliably point to best practice construction performance. While the emerging proponents of modern construction technologies and methods claim individual performance achievement, these are often selective and self-serving. Most research commissioned in this area is for internal corporate consumption, selective use and marketing. These initiatives normally cease when their sponsor loses interest.

It is argued that the use of non-price measures for construction performance has enormous potential when specifically targeted at creating common leading edge indicators to better inform strategic action and recognition. These measures can guide the way forward to a more productive industry, while better informing strategic knowledge and innovation. By comparison, construction-price measures tend to act as trailing indicators.

The important benefit of using non-price measures is that they uncouple the commercial sensitivities of contractors sharing their pricing details. The use of non-price measures enables public and private clients to accumulate simple access to important productivity insights that will lead to improving the future effectiveness of their project delivery. These measures will add to the pre-contract information that clients have to select their pre-construction (mainly consultant) teams, their construction delivery method and constructors.

Within five years the assembly and tracking of these measures will add evidence-based pre-qualification tools that both the industry's customers and contractors may use to benchmark the performance improvement trajectories they have adopted. The benefits of achieved improvement will translate into better value for money outcomes and enable new informed innovation to be targeted. They will provide a basis for first movers who adopt these tools to differentiate their offerings from those still locked into construction's past.

The performance of construction projects is quickly becoming similar to that on the factory floor in industrialised industries where the effectiveness of the prior design, planning, procurement and management systems are demonstrated and played out. As more construction fabrication moves off-site, the more real this analogy becomes. These trends will define construction's future and the success of its many players.

References and further reading

Abdel-Wahab, M. and Vogl, B. (2011) Trends of productivity growth in the construction industry across Europe, US and Japan. *Construction Management and Economics*, **29** (6), 635–644.

Agarwal, R., Chandrasekaran, S. and Sridhar, M. (2016) *Imaging Construction's Digital Future*. McKinsey and Company. www.mckinsey.com/industries/capital-projects-and-infrastructure/our-insights/imagining-constructions-digital-future.

Ashworth, A. and Perera, S. (2015) *Cost Studies of Buildings*, 6th ed. (London: Routledge).

Best, R. (2010) Using purchasing power parity to assess construction productivity. *Australasian Journal of Construction Economics and Building*, **10** (4), 1–10.

Blismas, N., Pasquire, C. and Gibb, A. (2006) Benefit evaluation for off-site production in construction. *Construction Management and Economics*, **24** (2), 121–130.

Chan, T. K. (2009) Measuring performance of the Malaysian construction industry. *Construction Management and Economics*, **27** (12), 1231–1244.

Chan, T. K. (2011) *Comparison of Precast Construction Costs: Case Studies in Australia and Malaysia*. In Proceedings of 27th Annual Conference of the Association of Researchers in Construction Management, ARCOM 2011, Bristol, **1**, 3–12.

Chan, T. K. and Hiap, P. T. (2012) A balanced scorecard approach to measuring industry performance. *Journal of Construction in Developing Countries*, **17** (1), 23–41.

Chancellor, W. (2015) Drivers of productivity: A case study of the Australian construction industry. *Construction Economics and Building*, **15** (3), 85–97.

Chancellor, W. and Abbott, M. (2015) The Australian construction industry: Is the shadow economy distorting productivity? *Construction Management and Economics*, **33** (3), 176–186.

Cherrafi, A., Elfezazi, S., Chiarini, A., Mokhlis, A. and Benhida, K. (2016) The integration of lean manufacturing, Six Sigma and sustainability: A literature review and future research directions for developing a specific model. *Journal of Cleaner Production*, **139**, 828–846.

Cho, A. and Parsons, J. (2015) Transportation industry faces digital 'disruptions'. *ENR: Engineering News-Record*, **274** (2), 1.

Christensen, C. M. (1992) *The Innovator's Challenge: Understanding the Influence of Market Environment on Processes of Technology Development in the Rigid Disk Drive Industry* (Cambridge, MA: DBA thesis, Harvard University Press).

Christensen, C. M. and Tedlow, R. S. (2000) Patterns of disruption in retailing. *Harvard Business Review*, **78**, 42–45.

Christensen, C. M., Verlinden, M. and Westerman, G. (2002) Disruption, disintegration and the dissipation of differentiability. *Industrial and Corporate Change*, **11** (5), 955–993.

Čiutienė, R. and Thattakath, E. W. (2014) Influence of dynamic capabilities in creating disruptive innovation. *Economics and Business*, **26**, 15–21.

Clark-Reynolds, M. and Pelosi, A. (2016) *When Did Disruption Become a Good Thing?* BRANZ. www.branz.co.nz/cms_show_download.php?id=0e68fa0ec0c524c957cadb790c0dd58d3eb95c07.

Crawford, P. and Vogl, B. (2006) Measuring productivity in the construction industry. *Building Research and Information*, **34** (3), 208–219.

de Valence, G. (2001) Defining and industry: What is the size and scope of the Australian building and construction industry? *Australian Journal of Construction Economics and Building*, **1** (1), 53–65.

Doloi, H. (2008) Application of AHP in improving construction productivity from a management perspective. *Construction Management and Economics*, **26** (8), 841–854.

Dubois, A. and Gadde, L. E. (2002) The construction industry as a loosely coupled system: Implications for productivity and innovation. *Construction Management and Economics*, **20** (7), 621–631.

Egan, J. (2002) *Rethinking Construction* (UK: Constructing Excellence). http://constructingexcellence.org.uk/wp-content/uploads/2014/10/rethinking_construction_report.pdf.

Foulkes, A. and Ruddock, L. (2007) Defining the scope of the construction sector. In: *Proceedings of the 8th IPGR Conference*, Salford.

Gambatese, J. A. and Hallowell, M. (2011) Enabling and measuring innovation in the construction industry. *Construction Management and Economics*, **29** (6), 553–567.

Gibb, A. G. F. and Isack, F. (2003) Re-engineering through pre-assembly: Client expectations and drivers. *Building Research and Information*, **31** (2), 146–160.

Glenigan. (2015) *UK Industry Performance Report* (Glenigan, Constructing Excellence, CITB). www.glenigan.com/sites/default/files/UK_Industry_Performance_Report_2015_883.pdf.

Guerrini, A., Martini, M. and Campedelli, B. (2013) Measuring the efficiency of the Italian construction industry. *International Journal of Business Performance Management*, **14** (3), 307–325.

Hardie, M. (2016) Vectors of technical innovation delivery by small and medium Australian construction firms. *Construction Economics and Building*, **16** (3), 59–70.

HM Government (2013) *Construction 2025 (Industrial Strategy: Government and Industry in Partnership)*. Her Majesty's Government. www.gov.uk/government/publications/construction-2025-strategy.

Holt, G. D. and Goulding, J. S. (2016) Positioning construction businesses on an 'evolution–innovation' continuum: Conceptualization of the 'equivocal zone'. *International Journal of Construction Management*, **16** (3), 220–233.

Holt, G. D., Goulding, J. S. and Akintoye, A. (2016) Enablers, challenges and relationships between research impact and theory generation. *Engineering, Construction and Architectural Management*, **23** (1), 20–39.

Jang, S-W. (2013) Seven disruptive innovations for future industries. *SERI Quarterly*, **6** (3), 94–98.

Jarkas, A. (2015) Effect of buildability on labour productivity: A practical quantification approach. *Journal of Construction Engineering and Management*, **142** (2), 1–5.

Karimi, J. and Walter, Z. (2015) The role of dynamic capabilities in responding to digital disruption: A factor-based study of the newspaper industry. *Journal of Management Information Systems*, **32** (1), 39–81.

Kenley, R. (2014) Productivity improvement in the construction process. *Construction Management and Economics*, **32** (6), 489–494.

Landers, J. A. Y. (2015) 3-D printed building highlights promise of innovative materials and methods. *Civil Engineering*, **85** (12), 29–31.

Langston, C. (2013) Development of generic key performance indicators for PMBOK using a 3D project integration model. *Australasian Journal of Construction Economics and Building*, **13** (4), 78–91.

Langston, C. (2014) Construction efficiency: A tale of two developed countries. *Engineering, Construction and Architectural Management*, **21** (3), 320–335.

LCI (2017) *Lean Construction Institute*. www.leanconstruction.org/.

Martínez, S., Jardón, A., Víctores, J. G. and Balaguer, C. (2013) Flexible field factory for construction industry. *Assembly Automation*, **33** (2), 175–183.

McKinsey & Company. (2016) *The CEO Guide to Customer Experience*. McKinsey & Company. www.mckinsey.com/business-functions/operations/our-insights/the-ceo-guide-to-customer-experience?cid=other-eml-alt-mkq-mck-oth-1608.

Miller, R., Strombom, D., Iammarino, M. and Black, B. (2009) *The Commercial Real Estate Revolution: Nine Transforming Keys to Lowering Costs, Cutting Waste and Driving Change in a Broken Industry* (New York: John Wiley & Sons).

Moreau, F. (2013) The disruptive nature of digitization: The case of the recorded music industry. *International Journal of Arts Management*, **15** (2), 18–31.

Nasir, H., Ahmed, H., Haas, C. and Goodrum, P. M. (2014) An analysis of construction productivity differences between Canada and the United States. *Construction Management and Economics*, **32** (6), 595–607.

Newton, C., Wilks, S., Hes, D., Aibinu, A., Crawford, R. H., Goodwin, K., Jensen, C., Chambers, D., Chan, T. K. and Aye, L. (2012) More than a survey: An interdisciplinary post-occupancy tracking of BER schools. *Architectural Science Review*, **55** (3), 196–205.

Ofori, G. (2001) Indicators for measuring construction industry development in developing countries. *Building Research and Information*, **29** (1), 40–50.

Pimentel, K. (2010) The rendering revolution: A desired disruption. *Computer Graphics World*, **33** (2). www.cgw.com/Publications/CGW/2010/Volume-33-Issue-2-Feb-2010-/Viewpoint.aspx.

Rojas, E. and Aramvareekul, P. (2003) Is construction labour productivity really declining? *Journal of Construction Engineering and Management*, **129** (1), 41–46.

Safe Work Australia (2015) *Work-Related Injuries and Fatalities in Construction, Australia, 2003–2013*. www.safeworkaustralia.gov.au/system/files/documents/1702/fatalities-in-construction.pdf.

Salmon, J. L. (2012) Wicked problems in construction, a new legal framework is required for the built industry to move forward effectively with BIM and IPD. *AUGIWorld Magazine*, February.

Sezer, A. A. and Bröchner, J. (2014) The construction productivity debate and the measurement of service qualities. *Construction Management and Economics*, **32** (6), 565–574.

Tam, V. W. Y., Tam, C. M., Zeng, S. X. and Ng, W. C. Y. (2007) Towards adoption of prefabrication in construction. *Building and Environment*, **42** (10), 3642–3654.

Vereen, S., Rasdorf, W. and Hummer, J. (2016) Development and comparative analysis of construction industry labour productivity metrics. *Journal of Construction Engineering and Management*, **142** (7), 1–9.

Vogl, B. and Abdel-Wahab, M. (2015) Measuring the construction industry's productivity performance: Critique of international productivity comparisons at industry level. *Journal of Construction Engineering and Management*, **141** (4).

Willis, C. J. and Rankin, J. H. (2012) The construction industry macro maturity model (CIM3): Theoretical underpinnings. *International Journal of Productivity and Performance Management*, **61** (4), 382–402.

Yi, W. and Chan, A. (2014) Critical review of labour productivity research in construction journals. *Journal of Management in Engineering*, **30** (2), 214–225.

Yu, H., Al-Hussein, M., Al-Jibouri, S. and Telyas, A. (2013) Lean transformation in a modular building company: A case for implementation. *Journal of Management in Engineering*, **29** (1), 103–111.

Index

Note: page numbers in *italic* indicate a figure and page numbers in **bold** indicate a table on the corresponding page.